T0302136

Essential Guide to Toolbox Talks

A Toolbox Talk (TBT) is a pre-job task made to ensure all parties involved in a task have full understanding of what they should do. TBTs are not easy; however, the author hopes this book will provide the right guidance to make TBTs less daunting, and easier to use. They are not a tick-the-box exercise; therefore, please do not treat them as such. This book challenges this perception and puts forth a case to consider TBTs essential to delivering safe working environments, thereby providing a complete understanding of the task.

Essential Guide to Toolbox Talks focuses on improving three Non-Technical Skills (NTS) related to TBTs: Communication, Situation Awareness and Stress Management. With communication the book looks at the impact of inappropriate questions, allowing time for questions/feedback, use of familiarities and knowing the answer and the errors that can occur. For situation awareness, the author conveys that effective communication enhances the environment in which people work, checking that the group understands the tasks; further coverage of what to do when things go wrong underpins this. Finally, it covers stress management and how important it is to confirm people know what to look for during a TBT and what to do if they believe a team member is under stress. With the onus on these three vital NTS, the reader will be able to deliver focused and much-improved TBTs that guarantee safe and effective performances in the workplace.

This neat and punchy book will change that tick-the-box perspective of the TBT. It will appeal to both practitioners in Human Factors and anyone at the front line in high-risk industries where TBTs are a requirement, especially those in oil and gas, nuclear, construction, logistics, transport and aviation who will be certain to consider it an essential guide.

Essential Guide to Toolbox Talks

Banishing Boredom and Re-thinking the Routine

Scott Moffat

CRC Press
Taylor & Francis Group
Boca Raton London New York

CRC Press is an imprint of the
Taylor & Francis Group, an **informa** business

Designed cover image: Summit Art Creations/ShutterStock

First edition published 2025
by CRC Press
2385 NW Executive Center Drive, Suite 320, Boca Raton FL 33431

and by CRC Press
4 Park Square, Milton Park, Abingdon, Oxon, OX14 4RN

CRC Press is an imprint of Taylor & Francis Group, LLC

ISBN: 9781032796284 (hbk)
ISBN: 9781032784946 (pbk)
ISBN: 9781003493105 (ebk)

DOI: 10.1201/9781003493105

Typeset in Times New Roman
by Newgen Publishing UK

Contents

Acknowledgements..vii
Author Bio ix
Acronyms/Offshore Phrases/Scottish Slang/Words......................................xi

Chapter 1 Introduction 1

Who Am I, or More Importantly Who/What Am I Not?..................... 1

Chapter 2 Introduction to NTS 13

Six NTS 13
Which of These Six NTS Do You Think You Will Be
    Good At? 14
"Which of the NTS Do You Think You Do Not Need
    to Know Anything About?"... 16
"Why Are You Only Examining Three NTS?" 17
"What Do You Know About TBTs?" 17

Chapter 3 Communication 19

Inappropriate Questions 20
Leading Questions... 26
Knowing the Answer.. 28
Allowing Time for Questions and Feedback...................... 30
Define What Is Meant by the Term Communication.......... 30
Use of Familiarities 31
We Can Read Each Other's Minds..................................... 32
Get Everyone Involved.. 32
Communication – Must Do's 33

Chapter 4 Situation Awareness.. 34

What Is Situation Awareness? 34
Memory 35
How Often Do You Do Your Offshore Survival?.............. 41
Distractions Task 43
In Theory, Should You Remember/Process More Numbers
    or Words?.. 45
Situation Awareness – Must Do's...................................... 47

Chapter 5   Stress Management..48

What Is Stress and Why Is It so Difficult to Define? 48
Different Types of Stress 50
There Are Four Main Ways to Identify Acute Stress 53
Stress Management – Must Do's...59

Chapter 6   Summary of All Key Points from All Sections 61

Index..63

Acknowledgements

I might be the author of this book; however, I most certainly did not achieve this on my own. I would like to thank everyone who has helped me over the years, whether that is professionally or personally. I would not be where I am today or have the knowledge and ability to write this book if it were not for everyone who helped/believed in me and allowed me to be the person I am today. I never forget I am only able to write this book as I have an understanding of offshore living and working, something I would never have managed with starting out as a tank cleaner, which was not the nicest of jobs; I was not very good at it but I enjoyed it and took a lot from it. A wise man, who is sadly no longer with us, once told me:

> Never think anyone is better than you or you are better than anyone else. Always take pride in what you do and give everything you have, no one can ever ask anymore. If it doesn't work out, then it doesn't matter at least we have tried. We all learn by failing, what is the worst that can happen, give it a go.

So I took this advice and wrote this book, I hope you enjoy and more importantly take something away from it on how to improve the effectiveness of a Toolbox Talk.

Author Bio

Scott Moffat, having a varied background since working offshore, is one of the leading experts within the NTS discipline. To date, he has observed over 100 groups in simulation exercises, over 90 external simulations and over 50 offshore NTS observations. All of this data plus his working background makes Scott feel just as at home in the simulator as he does offshore and thrives in sharing this knowledge with new groups.

Although he has a strong understanding of all the NTS – his area of interest lies in helping those on the front line understand and apply Communication, Situation Awareness and Stress Management.

Recently Scott has developed and facilitated a unique programme of TBT coaching across the Oil and Gas Industry building on his own experience offshore, in the classroom and through the observation of hundreds of TBTs over the years. Scott's unique ability to blend theory with context and humour make this vital reading for anyone looking to improve their knowledge or understanding of TBTs.

Acronyms/Offshore Phrases/ Scottish Slang/Words

These definitions are based on commonly accepted meaning in the working environment/wider outside world.

2/3 or 3/3	this means the amount of time for an offshore rotation, i.e. amount of time offshore then onshore.
Arc flash suit	specialised protective clothing used mainly by the electrical team when working on/near live electrical equipment.
Ballistic	being angry or upset.
Black Start Procedure (BSP)	a procedure which walks you through the processing of getting the lights back on if you were to go black (no lights on).
Bunk bed	Yes, it tends to be the traditional bunk beds. Never nice when you are on the top bunk. I am 16 stone, give or take, so me trying to get up to the top bunk is some sight I tell you.
Clambering	climb or move into a space.
Cock-a-hoop	happy/delighted.
Company person	the person on the rig site who represents the company who own the rig.
Confirmation bias	see what you are expecting to see (if you do a task 100 times by the time you come to do it for the 101st time your brain could have a preconceived idea of what it is expecting to see).
Doddle	simple or easy task.
Dog house	in drilling terms, it tends to be a shelter where the driller tends to operate from.
Doing my nut in	something that is annoying you. Nut being head.
Down hole	putting equipment down into the well (oil and gas reservoir).
Drill floor	section of the rig where, shockingly, the drilling is controlled and constructed.
Drivelling on	talking rubbish/nonsense.
Eejit	different way of saying idiot.
Elec techs	part of the electrical team.

Faffing	doing something in a disorganised way.
Flappin	panicking or stressing.
Fogged on	when the fog is so thick the helicopter cannot land on the rig. You are stuck there until it clears.
Foochery	fiddly.
Gaffer	supervisor.
Gist	main point/idea/theme.
Goodun	good thing/person/experience.
Hanging	rough/suffering from a hangover.
Herself	the better half, girlfriend or wife.
HF	human factors.
High-voltage (HV) isolation	where the electrical team isolate the high-voltage current to ensure they can work on it safely. How they do this I have no idea.
Inbetweenie	having an unscheduled hot beverage, i.e. cup of tea or coffee, whilst meant to be working.
Jobbie	can either mean faeces or in some places in Scotland a task or job (I feel I know what I was being referred to).
Jobs a goodun	the job/task you are away to take part will be completed successfully.
Kick seven shades	to use aggressive means on other people (this phase normally has a swear word at the end, I decided to leave it out).
Kinda	shortened version of kind of.
Malarky	nonsense, rubbish. Sometimes we Scots use the term to describe something in general, and does not necessarily mean we think it is rubbish.
Mechies	mechanical team.
Medevac	medical evacuation from offshore.
Mental	when someone is mad, tends to end in a party getting shouted at.
North Sea Tigers	absolutely no idea where this term came from; however, it is commonly used when describing people who work in the oil and gas industry (especially those who work offshore).
Nth degree	to the upmost/maximum, i.e. I have used my computer to the nth degree.
NTS	non-technical skills.
O+G	oil and gas industry.

OIM	Offshore Installation Manager (highest ranking person offshore).
Park	out on site (there are no slides and roundabouts etc. offshore (sadly)).
Peely-wally	looking ill, tired or drained.
Porkies	little white lies.
Raging	angry or annoyed.
Reckon	think, i.e. I reckon I could go and climb that tree.
Rejig	reorganise, rearrange or go back to the start.
REP/s (or whatever way around some companies have their letters)	Responsible Electrical Person who is in charge of all electrical work offshore.
Revving	putting your foot on the accelerator whilst normally in neutral.
Ringing	soaking wet.
Self-fulfilling prophecy	behaving in a manner or working towards that behaviour that has been labelled towards you, i.e. Scott is the joker of the group.
Shat	vulgar slang for going to the toilet that rhymes with hit.
Shifty	behaving in a weird or bizarre manner.
Shinty	traditional Scottish (mainly Highlands and West Coast) sport. The easiest way to describe might be ice-hockey on grass.
Sim	simulator.
Snot	nasal mucus.
Sparky/sparkies	electrical people or electrical team.
Taking the mick	having a joke with others/ridiculing each other.
TBT	Toolbox Talk (hopefully everyone will have a much better understanding of what these are and how to do a more efficient TBT by the end of the book).
Tech stuff	technical skills/knowledge/ability.
Timber	to have put on a little timber, to have put on a little weight.
Wanna	want.
Wee	small.
Willy-nilly	to throw something around, i.e. an idea or opinion in a disorganised/unplanned manner.
Ya	you.
Yer man	that person.

1 Introduction

WHO AM I, OR MORE IMPORTANTLY WHO/WHAT AM I NOT?

"Are you the psychologist are you?" asked the naked man sitting on the bottom bunk of the Offshore cabin when I walked in for the first time. "I am not, but if I were I would have a field day with you".

I am not a psychologist, nor I am a doctor or professor. I am pretty sure a couple of lines into this book you would have soon realised this. Writing a book, which can be deemed quite academical, has never really appeared on my radar. I have helped write chapter/s in other books and journal papers; however, this will be the first time I have been let off the lead. I might not be the most academic; however, I have quite an eclectic working background which has allowed me to develop certain skills, not as exciting as the ones Liam Neeson once described, which I can contextualise to ensure the theory I present can be easier to process.

Work Experience

Every job I have done in the past has contributed to where I am today (Human Factors Director) and has allowed me to have a greater understanding/application for my main area of expertise: the six Non-Technical Skills (NTS) (Stress Management, Communication, Situation Awareness, Decision Making, Teamwork and Leadership). I must add the caveat that just because these previous jobs have allowed me to be where I am today does not necessarily mean I was any good at them; for instance, when I worked, briefly, as a stacker in the wood I am pretty sure someone said I was "About as much use as a man short". This job along with a few others were jobs I had before I went to University and probably did not really aid in my career development (well maybe being a voluntary Firefighter, but again I would fall under the useless category).

On to the jobs, I feel these experiences have contributed to me to be where I am today.

DOI: 10.1201/9781003493105-1

CARE ASSISTANT AT A SPECIAL NEEDS SCHOOL

During my days at University, I spent most of my time away from Uni as a care assistant at a special needs school. It started as just a way to earn some pocket money; however, it soon turned into one of the best jobs I have had. It allowed me to understand people from a different perspective and also let me see how people communicate and process information completely different. I did this all in all for around 4 years and in the final year I was one to one with a student. This role was basically me trying to help this person into society, i.e. taking them to work, getting to and from their home. Basically teaching them general life skills, to be honest at times I was not sure who was teaching who as there was one instance when we finished his work and we went to the cinema, a reward they chose. As we were running late, I made the decision for me to run and get the tickets and for the person to wait near in the large open area in between myself and the pick n mix. At this stage, I might add this person had pica, which basically means said person will eat anything edible or not, for example, it was not uncommon for this person to eat the insoles of their shoes or even frozen puff pastry. As I stand next in line I soon realised I might have made a grave mistake leaving them where I did (as I feel the pick n mix was looking very appealing). I might also add his chosen outfit for the day was a t-shirt (which read I heart Canada) and trousers which if I ever wore to school might have been referred to as the floods were coming (they were way too short). Finally, when this person got excited, standing very near the pick n mix, they would make a loud noise not to dissimilar to the revving of a car. As I start to state the film, we wanted to watch I could hear their excitement levels going through the roof, "Two tickets to see Harry Potter please" or whatever the film was. At the time there was a discount for special needs students, so I said "sorry special needs student and one ticket for their carer". Now let's bear in mind the clothing choice and noise coming from the person I was with; it then came as bit of a shock when they said "Which one is the carer?" Trying not to laugh but also feel a little insulted I said: "Do you think that person who is about to destroy your pick n mix is in charge of me"?

To say I genuinely loved this job would be the understatement of the year. It was the first job where I actually felt not only was I helping people, but I was actually not too bad at it. If there were any "about as much use as man short" comments, then I certainly never heard them. This job had such an impact on my life I thought about trying to get into educational psychology; however, due to circumstances at the school and in my own life, this route was never explored and, to be honest, once I really knew where I was going I have never looked back, I have never forgotten my time at that school.

OFFSHORE

The next stage in my career was going offshore to be a tank cleaner. This came about as my friend knew my time at the school was coming to an end and asked if I wanted to go offshore. I remember saying to them that I did not have any qualifications to do this job and he replied "you do not have a criminal record, so you are doing better than most".

This was my introduction to the Oil and Gas Industry, I would love to say I took to it like a duck to water. However, this was not the case; it took me a good while to get used to working in quite a harsh environment for up to 3 weeks at a time (in the North Sea, this can increase if working overseas) normally on a 3/3 rotation. There are numerous combinations for rotations: 2/2, 2/3, 2/3 2/4, 3/3, 3/4 3/5. Other than trips offshore with my current company, I have never done less than 3 weeks offshore. Once, I thought I was only doing 2 weeks. I left the HQ of the company I was working with at the time and the gaffer said

> Right then Moffat, as this is such a big job and the bosses are happy we with the progress they have decided to pick one person at random who will only be working 2 weeks. It will be a different person each trip and you my friend have been selected for this trip.

I am now sitting in the car absolutely cock-a-hoop, and for the first time in my off-shore life, I was looking forward to getting offshore. The difference between working 2 weeks compared to 3 weeks offshore is huge; it might not sound as such but that extra week is a killer. Now after about 11 days off this trip offshore I am jumping about like Tiger from Winnie the Pooh with a smile, a wire brush and Dettol would not remove. The gaffer comes up to me and says "What are you jumping about for" "Well I am going home in a couple of days" to my amazement, he then burst out laughing and said something along the lines of "Are you really that gullible? Do you really think you are going home after 2 weeks? My god Moffat this is a new low, even for you" I am now standing there, probably with my bottom lip nearly on the floor, and drying my eyes. "Get your arse down that tank" the last 10 days were the longest 10 days I have ever endured offshore. The difference in doing 2 weeks, well when I thought I was doing 2 weeks, was unbelievable once you get through the first weekend you are on the countdown for home, takes a bit longer to get that feeling for 3 weeks.

Although cleaning tanks was not the most pleasant jobs in the world, I learnt a lot during my first 1.5 years offshore. I was introduced to some of the best leaders I have ever met; to be fair I also worked with some of the worst leaders I have ever met. The basics of leadership I learnt during my stint cleaning tanks, I still try to apply in my working life now.

I honestly do believe if I did not do this role, I would not be able to teach the Toolbox Talk (TBT) training sessions offshore today. It allowed me amongst other things to contextualise the theory to ensure everyone gains as much as they can. I feel it is always easier to process information with it is put into context. Granted not everyone offshore has been down a tank, but the general working and living offshore is the same.

From here I went on to become a Coring Engineer, which is basically throwing, not literally, equipment down the hole into the reservoir to see if there is any oil or gas (granted it is a bit more complex than this but that is the general gist). The job is based on the drill floor, so although I do not understand the technical aspects of drilling I have been around many drilling teams, therefore I understand the team dynamics of a drilling team. It was this job that made me feel so comfortable when teaching NTS to drilling teams, as again I was able to contextualise the theory to ensure they had a better understanding.

This job almost did not happen for me, as I walked out whilst waiting for my inter-view. Still to this day, I must thank a random taxi driver for getting me where I am today. I was waiting in the reception of the Coring company, and I was listening to two other people, who I assumed were other candidates, talking about where they got their Master's and what they planned to do. I am now sitting there thinking I work down a tank maybe this position was a step too far or not really for me. I go to the receptionist and politely say I would like to decline the interview and leave. "Are you absolutely sure you want to leave" "I am really sorry for any inconvenience caused and really sorry for any hassle I have caused. But yes please I would like to leave". I slowly walk out of the main building and head for the taxi. I open the door and slouch myself into the back seat of the taxi and really just wanting to get on the road and get home. However, the taxi driver had other plans in that he sat there and said, "What are you doing here?" "Listen man, I am not really in the mood. I just wanna go home. So if you do not mind, please just take me home". To my surprise he reached up and turned off the meter, "Erm what are you doing?" "Turning off the meter, what does it look like I am doing?" I was not really in the mood for smugness here, so replied with something along the line of "Look man I am not entirely sure what you are trying to do here but there are people in there talking about their MSc and what they have done before. Trust me when I say it well and truly trumps me, so I know you mean well but please take me home". "What have you got to lose?" "What is the worst that could happen? I am not trying to be your Dad here". "Really cos where I am sitting that looks like exactly what you are trying to do". "Why don't you go back in there say you are sorry and just see what happens and I promise I will sit here the entire time with the meter off" "I totally get what you are saying, I am tired and I just wanna go home, so please can you just take me home".

Look you have one shot in this life, so why don't you head back in and see how it goes. Yes it will be a little embarrassing, however I have no intention of moving so you can get a little embarrassed when you walk back in or we can sit here for another couple of hours until people start to leave from here and go home and low and behold we are still sitting here, which do you think is the most embarrassing?

"Sitting here when they leave will not bother me, I feel it might be a little different for you" "Ok, Ok I'll go in". Now feeling pretty disgruntled I go to leave; however, to show just how annoyed I was I went to slam the back slidey door; however, unbe-known to me it was one of these self-closing doors. I am now standing there waiting for the door to slowly close. I apologise and made my way back into the building.

As it turns out I ended up getting the job and when I got into the taxi for the second time and told the driver he genuinely looked pleased for me. As I go to leave the taxi, after trying to give him more money that the meter indicated. I remember him saying "If you are ever in a similar position in the future then could you do the same. No matter what it is, it is always better to try and have a gone than give up at the first hurdle". Without this taxi driver, I genuinely do not think I would be where

I am today, this job was all about working on the drill floor offshore. This experience allowed me to gain an understanding of how things work on the drill floor and therefore contextualise the NTS theory to these teams. To date, I have observed around 80 drilling teams in a drilling simulator. An experience I would not have had if the taxi driver had taken me home. If by slim chance you are reading this, then thank you Mr taxi driver, literally could not have done it without you.

LECTURER IN PSYCHOLOGY AND SOCIOLOGY

After pretending to be a North Sea Tiger for a while, I was a teacher/lecturer in psychology and sociology for 2 years. I taught a range of topics (stress, memory, research methods to name but a few) to a range of different levels from Intermediate 2 up to Higher National Diploma level, apologies I am not sure what these levels equate to nowadays, nowadays alright granda!!

I was absolutely bursting with pride to have been given this opportunity and could not wait to get started. I remember thinking back to my college/Uni days where the lectures I enjoyed were the one where the lecturer could bring in real-life experiences/ stories and had a natural desire and enthusiasm about the topic and this was all that was required to teach. A quick self-assessment before I enter my first classroom yip, I have these two skills in abundance, so I am going to go in here and become the lecture the students had always wanted. In my naivety I had failed to realise that yes these two skills were incredibly important; however, becoming a successful teacher/ lecturer was a hell of a lot more than this, I very quickly found this out (the hard way obviously).

The first time in front of the class all I had to do was stand up and introduce myself (which was around 20–25 students ranging in age from 16 to 18 years old) who I would be teaching all on my own in less than a week, I mean how difficult could this be? I found out it was very difficult in the next 20 seconds. I hear the Head of Department from the College say "I will be taking the class today however before we begin I will hand you over to your new psychology lecturer who will be taking you for the next year". Here it is my time to shine, my time to teach the next generation of psychology students and this introduction will be nowhere near enough as I am full of positivity and passion to deliver this topic, let's go.

As I stepped forward to introduce myself, my legs suddenly felt incredibly heavy, then the basic function of saying my name turned into the one of the most difficult functions imaginable. As I opened my mouth to say the simple sentence: "Good morning everyone I am Scott Moffat and I will be your new psychology lecturer … . Enjoy the class today and I will see you next week", jobs a goodun sit down and relax. However, this sentence was replaced by nothing, well not completely nothing a bizarre sound, like a wee squeak came out (to be honest I would have preferred nothing). Even I thought, what on earth was that noise. This weird squeaking noise prompted the first laugh, of many I might add, from this class. I am standing in front of a class of young people who were sharp as a tac and instead of giving me a minute to compose myself one of them shouted "speak up ya eejit".

I wish I could say that upon hearing this slur being thrown in my direction, I was hoping the rest of the class would turn on this individual and say along the lines of: "that is well out of order give that man a chance… . Sorry Mr Moffat you may proceed"; however, this did not quite happen. Upon seeing the reaction this one pupil received, laughter, more pupils thought it would be a great idea to join in with either sniggers or their own lovely terms of endearment. Now instead of: that was easy/job is a goodun and sit down and think positively about my new chosen career, I was being verbally abused by a good proportion of the class, great start.

After seeing this absolute train wreck of an introduction, the Head of Department quickly took control of the situation and the class. My main job and focus were to sit there for a full two hours to try and understand what just happened and what the hell I was going to do next week when I have to deliver a two-hour lecture on developmental psychology on my own.

The deluded thought of myself being the catalyst in helping these pupils start their passion and love for psychology was dwindling by the second and leaving that well-paid offshore job was fast looking like a massive mistake.

Over the next 3 weeks, it did not get much better for me or for those poor students who had to endure my "painful" classes. However, much of this was my own doing. I became incredibly fixated on the product which was to finish the class as soon as possible and not the process in between, i.e. the actual teaching and building up a rapport with the students and ultimately given them the learning environment they deserved. I was beginning to live up to the expectations of the students, self-fulfilling prophecy at its finest, and my classes were actually getting worse, which I didn't think was even possible.

The turning point in the class and ultimately my career came by sheer fluke. I was standing in front of the class reading every word from my meticulous notes which I had written down on quality A4 paper, if you have ever seen the Friends episode where Ross practises his first lecture to the rest of the group, well I was 10x worse than that, over exaggerated eye contact and all. I was completely aware that not one pupil was playing the slightest bit of interest to what I was saying; however, I did not want to stop as the fear of being ridiculed (again) was far greater than me completely embarrassing myself which I was currently doing. The pupils were either talking amongst themselves or doing work for other classes, the classes they probably so wished they were attending. I turned around and realised, I had not changed the slides for a good while, so even if there was anyone listening to me drivelling on, they had no focal or reference point on the screen. For my own curiosity, I tried to find the appropriate slide, which took me a while, in fact I could not find it anywhere and soon I reached the end of my slides, where it said End of topic thank you, good luck with your revision. I looked at the slide and thought I'll just leave this on as let's be honest no one really cares about the slide, the topic, the class or even me. This was the sobering moment I knew I had completely failed and I had let so many people down: everyone at the college/school, myself but the one which really hurt me was the fact that I had let every single one of these pupils down, I found that really difficult to admit. It was time for me to stop putting everyone through this weekly misery. It was here I decided to call it a day, and after the class, I was going to hand in my notice and leave the school/college. Let's be honest it was probably for the best.

As I stand there, just wishing this experience to be over, my attention was drawn to one of the pupils, who unknown to them or even me for that matter was about to change both my teaching style and, to be a bit more dramatic yet accurate, my life:

"Sir can you go back as I didn't manage to get some of the notes"
"Firstly, thank you for putting up your hand I appreciate that and secondly, although we are in your school this is a college course so please call me Scott". I think a lot of them had actually forgotten my name.
When I finished this sentence I genuinely think everyone in the class at that time knew I had given up.
"Scott can you please go back".

It was at this stage that I noticed something very strange and it took me a minute to process what it was, some of the class were looking at me. More importantly maybe, just maybe, they were paying attention. I remember thinking if I lose their attention here, then it is well and truly over and forget it as I will never get it back. So I went for it.
"Who else did not manage to take down what was written on the slide?" Whatever the hell it was, God even I didn't recall what it was. One person who was sitting in a group of five, two of whom had actually turned their chairs around so they could hear the others better said "Sorry I didn't get it either". My response which was already half out before I could even think about it was along the lines of

well that is not my problem … we have been in here for over an hour and you have not tried to take any notes down, so as of right now if you have not written it down then tough, I have my qualifications you all do not. So if you want to learn and progress then you listen to what I have to say.

I was shaking like you would not believe but I was delighted I managed to get the words out. As I sat and let the words sink in, I was not doing this for effect, I genuinely could not believe I had just said what I had said, I must add I felt really bad for saying this however it had been said so let's see what happens. After what felt like the longest silence ever, one lad said how "how hell do you expect us to understand this and take notes when you read from your paper. It's really hard to hear what you are saying when you make it so hard to listen."
"Ok who else find it hard to take notes when I read from the paper?" Every hand in the class goes up, this did not bother me, well too much, for the first time since the head of the department took the class, all those weeks ago, I now had the attention of every single pupil. I was in unchartered territory, I put the notes down, I turned off the projector and I said "We are going to do something different, I will stand here and answer any questions you have on psychology, whether it is this topic or in general".
"It is not a free for all, it's not feeding time at the zoo" oh my word, did I actually just get a laugh, well a smile, from someone in the class? "One at a time, ask me any psychology related question you can think of". With in seconds at least three hands were in the air and not only that everyone was now moving to sit where they should be sitting, for the first time in weeks I now had what looked like a normal functioning

class, I certainly did not see this coming and I will not lie I was beginning to like it and the stress and anxiety of standing in front of the class was, slowly, beginning to leave me.

It was at this point I realised it was not my psychology knowledge, passion etc. which had led to disaster class after disaster class, it was how I was presenting my knowledge which was the major problem.

"Right first question let's go". I point to one of the students sitting to my immediate left, and now I realise just how bad a lecturer I was, as I didn't even know their name. What an idiot. I had been so self-obsessed with doing the class I had not taken the time to get to know any of the students. I managed to get away with it as me pointing in her general direction was all the encouragement she needed. The question "How did you get into psychology" came flying at me, this was one of the very few who took notes and actually looked like she was enjoying the class, not sure how but never mind.

It then donned on me I hadn't really introduced myself properly to the class. No ice breaker, no this is me now tell me about you, nothing just the A4 paper in front of my eyes and religiously read at them. Absolutely no interaction whatsoever, I am now surprised they even bothered turning up. Again, what an idiot.

Back to the question: How did I first get into psychology? I replied: "that is a great question to get us started" even as I said this sentence I could see what that small amount of praise had done to this girl, as she was grinning from ear to ear, this further emphasised just how much I had neglected these students.

"Strap yourselves in folks as this is quite the story" another little laugh, this can't be happening can it. Could I actually be getting the attention of the entire class?

"Who here has ever heard of Kitty Genovese?" "Who" KG was a young woman who was murdered on her way home from work in 1964. The ooohs and aaaaas from the class were now encouraging me, this is not the worst part, she was murdered in front of around 30 people, no idea if this was the correct number or not but I had their attention, and historical accuracies were the last thing on my mind. I continue, apparently it took over 20 minutes for someone to phone the police and when they finally arrived, they asked why no one phoned them sooner,

"Why do you think was the most common given answer?" this is me now asking the class:

"They didn't have mobile phone" true yes but more basic and more importantly more fundamental to that "They thought someone else was going to do it" said the same girl who asked the original question, who now looked like she was loving life. "Well done, that is it exactly it". It was here I first managed to actually engage my first audience, (do not get me wrong this was not the topic I was meant to be teaching but at this moment I didn't care in the slightest). I am not saying I went onto become the best teacher they ever had and motivated them to become the version of themselves. This was the start of the long process of me having the ability and confidence to facilitate information to an audience. If anyone reading this is ever thinking of getting into teaching/lecturing or presenting training courses then I would highly recommend it, however, please do not be as naive as me and please realise it might take a long time to facilitate to the level you might have seen from others. Once you get past the tough parts it really is the best job in the world. Even when I present to this day there is nothing else, I would rather be doing than standing presenting to people.

All these experiences now give me the knowledge and confidence to facilitate information on the NTS to a range of audiences from six people in a simulator training course to 30 people offshore when facilitating TBT training.

"So How Did Someone Who Used To Clean Tanks Go On To Do What You Do?"

This is a very good question, and it is one I ask myself on a regular basis and the simple answer is good old-fashioned nepotism. I say this is a jovial manner; however, the exact definition of nepotism might fit to my circumstances of getting the job as a Human Factors Advisor, now Director. However, I like to think my eclectic knowledge and background also played a major part. Basically, one of my best friends was sitting with his girlfriend's (now wife's) parents and they were sitting eating a Sunday roast, I am surmising here but I do not think it was too far from the truth. I like to think the conversation went along the lines of, "We are looking to expand the business, do you know of anyone who has a background in Psychology/Social Science and does not mind standing in front of people to deliver Psychology-based training". The second question I like to think was asked in a sarcastic tone "Has this person ever worked offshore?" "Well as a matter of fact my good friend Scott is currently working as a Psychology lecturer and spend about x amount of years offshore." "Well, this chap sounds like a perfect fit for our company, lets get him on board". I made the last sentence up; however, this is the general gist. So yes I got on the radar of the company owners via my friend; however, I like to think my years as a lecturer combined with my years offshore made me a perfect candidate for teaching the NTS to none other than offshore workers. The rest as they say is history. I am eternally grateful for both my friend and the owners for giving me the opportunity; I have never taken it for granted and I have given my all to ensure NTS are being integrated into the oil and gas industry.

So What Do I Do Now?

In short, I try to use all my previous work experiences to integrate Human Factors or NTS into companies in a variety of different industries. Due to my background, my main focus is integrating Human Factors or NTS into the Oil and Gas Industry. From here on the book will focus on integrating the NTS into industries, mainly the oil and gas industry.

Trying to integrate NTS, into the Oil and Gas Industry/any industry, can be challenging to say the least. As a result, I feel my offshore experience has given me the ability to put the information across in a way that I would have liked when I was offshore. Therefore easier for people to implement and describe to others when offshore.

One of the issues with the NTS is that they can be portrayed as the easy skills, common sense or the term that drives me insane is the soft skills. People do not realise that these skills require just as much training and practise as the technical skills.

We have been involved in conversations with people in the oil and gas industry when they are doing incident investigations, and we say: "What was the main cause of that incident/accident?" to which someone replied "Leadership". "What training have these people had in Leadership?" "Well, none, we do not need any as we are all technically very good". Oil and Gas and maybe other industries can be quite bad at making the massive assumption that just because someone is good at their technical stuff then they will automatically be great at skills like Leadership. Never make the assumption that just because someone is good technically they will automatically be good at leadership. Being an effective leader is a hell of a lot more than knowing your techy stuff, from a non-technical perspective, for instance, knowing and understanding what motivates individuals, coaching individuals and ensuring the team can work together to move forward to achieve the desired goal. I am summarizing here; I know there are more than these (the above and the NTS), but I am just making the point that there is a heck of a lot more that goes into Leadership than just techy stuff. I have met numerous REPs over the years, and some have been great technical, however, did not have the knowledge and skill to fully apply themselves as they did not have the NTS knowledge. I have also met some REPs who did not have the technical knowledge or ability, but that is neither here nor there for this book.

"Ok so Leadership was the main cause of that incident, what does it say was the main cause for the next incident?" "Well this one says poor Communication" "And what training have people in the Oil and Gas industry had on Communication?". "Well again none, don't need it do we, we can all talk away to each other and we are pretty good at that Communication malarky". Communication is a skill like any other, we cannot just expect people to go offshore, leap out of their bunk and be effective communicators. I will never run a marathon, if you saw me you would know exactly why, Sir Mo Farah I am not. Communication is like the skill of running, I cannot/would not just get up one morning and have the ability to run 26.1 miles, I reckon I would struggle to run 2.61 meters but hey ho. Communication, in fact all the NTS are exactly the same in that they take practise, skill and determination to be any good at them. Just because someone is good technically it does not mean they will be any good at the NTS. Combining technical knowledge, skill, ability with the NTS can reduce incidents and accidents which could have a positive impact, in my opinion, on safe and effective performance.

Almost Forgot: What Is a TBT?

In its most simple form, a TBT is a pre-job discussion which allows all party members to have a shared understanding of the task. It sounds pretty simple; however, they definitely are not, never underestimate the importance of the TBT.

In my opinion, the TBT is one of the most underrated and underperformed tasks that is completed in the oil and gas industry. This is certainly true within the Oil and Gas Industry from my own experiences. It can often be seen as a tick-the-box exercise and therefore may not be used to its full potential. This is the main reason I decided to write this book.

The main aims of the TBT are to address the following: define a clear and structured goal for the task, identify roles and responsibilities, identify risks and hazards, identify the 'what if' situation, identify any 'Stop the Job' triggers, undertake a worksite visit, and ensure individual and team verification of the tasks steps. The TBT does not have to be limited to these.

At this stage, I would like to emphasise that I know different companies in Oil and Gas might have a different way of doing things. In fact, different industries might also have a different way of doing things, as a result my aim is to not go through set procedures on how to carry out the TBT but using my knowledge of the NTS to give basic advice on how to improve TBT regardless of task, company or industry.

WHY DID I WRITE THIS BOOK?

I wrote this book not because I once pulled a muscle in my bum and could only sit on the chairs in our dining room table. We (the wife and I) were speaking to our friend who is an avid writer, amongst other things, and he gave me the best advise possible, he simply said just try and write it, do not overcomplicate or overthink it. Just sit down and write it, try and aim for 500 words a day, do not worry about the context or whether it makes sense just sit down and get used to writing. So I took this advice on board and whilst I was recovering from the pain in the buttock I started trying to hit my target of 500 a day. It took me a while to get going, but before I knew it I was hitting the target and some. I am not saying this process was easy but once I got into it and basically started writing in my style which is basically exactly how I talk, the process was not too bad, well the initial writing was not too bad and I struggled with the *foochery* bits, i.e. any part that I didn't have to use my HF experience, especially the Index, that was fun.

Over the next three chapters, I aim to discuss and break down Communication, Situation Awareness and Stress Management to enable people to get an understanding of how to conduct a TBT to enhance safe and effective behaviour.

STYLE OF WRITING THROUGHOUT THE BOOK: (BE YOURSELF)

I often use the term be yourself, whether that is facilitating a training course or delivering a speech. I decided to practise what I preach, in that I have decided to use a jokey/tongue-in-cheek approach, mainly because this is how I would facilitate a training course and, as a result, I have tried to do the same whilst writing. This style does not mean I think TBT or any other aspect of the NTS or HF is to be taken lightly or as a joke. I am incredibly passionate about what I do and I don't think for one second anything written in this book should be seen as a joke or taken lightly. However, I thought I would try and relay some of my stories and experiences as best I could. With regards to these stories, I have tried hard to keep them as close to reality as possible; however, with my aging brain they might not be 100% accurate but are to the best of my knowledge as accurate as I can remember. I have tried to keep the stories are true as possible and use the language that there were spoken in, like many high-hazard industries the Oil and Gas Industry is no walk in the park and out of respect for those telling the stories I tried to keep the language as realistic as possible. This book or the language used was never meant to offend anyone, so I apologise if this is the case. I understand some of the language used might be a little from the norm; however, I thought writing something like my supervisor said cor-blimey Moffat you are a silly boy would not quite work.

2 Introduction to NTS

SIX NTS

The six NTSs are Stress Management, Communication, Situation Awareness, Decision Making, Teamwork and Leadership, and they can be divided into three other categories:

- Cognitive skills – Situation Awareness + Decision Making
- Social skills – Teamwork, Leadership + Communication
- Personal recourse management – Stress Management (Flin et al, 2008).

When discussing the NTS I tend to use six, I know some people like to say seven, with the addition of coping with fatigue. However, I tend to use six and incorporate coping with fatigue into the other six skills.

Figure 1 was created in this way to highlight these NTS are dependent upon each other and that no one NTS is more important than the others. I know this might not represent the true relationship they have with each other, but I just wanted a nice easy diagram to reflect this.

Every single person gets stressed, and those who say they do not are telling porkies; stress can have an effect on Communication (in the sim/real life, it tends to reduce or stop altogether). If Communication has reduced, there is a chance Situation Awareness has reduced, which in turn can reduce the chances of an effective decision being made. If all these skills have gone, then Teamwork and Leadership might not be on anyone's radar. I know I have simplified this, but my point is that they are completely dependent upon each other. It is also important to note that these NTS are dependent on people's technical skills. I am not for a second suggesting that once you get to the end of this book or you have been on an NTS training course then you no longer need to use your technical skills. What I am saying is that they tend to go hand-in-hand to improve safe and effective performance. However, within the oil and gas industry, people tend to focus on the technical side of things and can neglect the NTS. They are both extremely important and both require training and practise.

A company approached us once and said we have had an incident on one of our assets and the problem looks like it was Communication. "Ok, what are you going to do about this? Are you going to train the person in Communication?" No, no

DOI: 10.1201/9781003493105-2

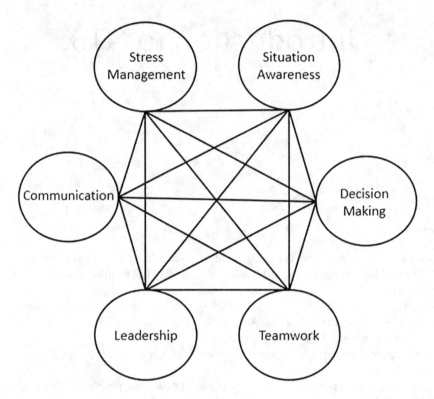

FIGURE 1   Non-technical skills diagram.

Communication is just common sense I am sure they will soon learn from it. They approached us again and said, "We have had another incident on the same asset and this time it was poor/lack of TBT was the main causal factor". "We can see a common denominator here (Communication), what are you going to do about it this time?" "Well, we might put them on a High Voltage/Low Voltage training course". *Please do not use a technical solution to try and solve an NTS problem.*

WHICH OF THESE SIX NTS DO YOU THINK YOU WILL BE GOOD AT?

The first question I tend to ask at the beginning of any training course, especially one involving a simulator: "When you are put under stress in the sim which of the six NTS do you think you will do well at?" This is an interesting question and during the one-to-one feedback session, I like to see which one/s they have written down. It is a good way to see what they think against what they were actually good at, perception versus reality. They are allowed to write down as many as they like; they must write at least one (we have only ever had one person who named all six, legend), he was not great at all six but fair play to him for giving it a go.

The most common given answer is Communication. "I have been talking since I was like 3 or 4 years of age what can yer man Moffat tell me about Communication?" Most people tend to be good at the talking/shouting aspect of Communication; however, it tends to be the listening aspect which lets them down. Listening is an incredibly important aspect of Communication (Flin et al, 2008). The groups who tend to do well in simulation exercises tend not to be technically any better than other groups; however, they tend to: listen to each other, challenge each other and then move forward as one group. The last simulation for our training course is aptly named Bad Day at the Office (BDATO). It is called BDATO due to the negative connotations it gives off; some people think they will walk into the simulator on a Friday morning and I will be standing there with two flame-throwers saying "OMG the rig is on fire, what are you going to do about it?" It is nothing like that, but throughout the week this seed can get planted of how stressful it will be during BDATO. Additional stress is applied just as the group are away to enter BDATO the words "you only have 1.5 hours to do this", couple that with the name BDATO and that is where most stress comes from, "Omg we only have 1.5 hours, and we only have 1 hour left, etc". We have had over 100 groups in the simulator for BDATO and only six have ever completed the exercise with no help at all. The answer to BDATO is always at least mentioned; it might not be discussed but it is always mentioned.

There was a recent example where during the introductions to a training course which are pretty bog standard who you are, what you do and how long you have been doing it, one person said they had spent a lot of time working abroad and therefore might not quite be up to the technical standard as the rest of the group; however, he was looking forward to learning from everyone on the course.

They were about half an hour into BDATO office and were at the tipping point, i.e. they could solve the issue in the next couple of minutes or they could go down a rabbit warren and not return when this person who had said he was not on the same level as them technically said "how about we try…" (I will not write the answer in case an eager REP reads it and smashes it within 5 minutes of entering the simulator), he basically said the answer; however, someone on the course said "what are you talking about you have already said that you do not have the same experience as us…" he tried again; "let's just try it and see what happens"; someone else piped up with "you heard what she said let us get on with it". We now think this could be interesting, he has given the answer, but no one is listening to him or even willing to try his suggestion. The next 10 minutes the group went down the aforementioned rabbit warren and we stopped the exercise 5 minutes later. We then went on, say why on earth did you not listen to what he was saying he had the answer and it would have saved a lot of pain (not actual physical pain), but it has now turned out to be one of the worst BDATO exercises we have seen, in fact, one of them even said: "Scott could have done a better job than us". Thanks for that I know I am not technical, but I do have feelings and I am standing right here.

If you do not listen to everyone in the group, whether that be during a simulator exercise or in real life, there is a chance you could not process an important piece of information which could lead to the group not having a full picture of the surroundings which could lead to less understanding and the group fragmenting. Listen to everyone, even the less experienced members of the group. Everyone has something they can bring to the group and sometimes those with less experience can identify things that other people could miss due to them doing it so many times. Everyone can talk; it takes skill and timing to listen to what people are saying.

Communication is the NTS which people think they will perform best at. It is also the one we spend most amount of time on for the classroom theory and it is the NTS I spend the most amount of time on for feedback (both onshore and offshore). Never ignore the importance of Communication and from that never ignore the importance of the TBT. In my eyes, your TBT is one of if not the most important jobs you will do daily.

Another reason I spend so much time on Communication is the fact that in the oil and gas industry, the loss of Piper Alpha (1988) Oil Platform along with the death of 167 workers was attributed to poor Communication (at shift handover), which was then compounded by leadership failings in emergency response (Cullen, 1990). Poor NTS are not only attributed to Piper Alpha or in fact the Oil and Gas Industry, for more details on other major incidents whereby poor or lack of NTS was a major causal factor (Flin et al, 2008).

"WHICH OF THE NTS DO YOU THINK YOU DO NOT NEED TO KNOW ANYTHING ABOUT?"

This is the second question I ask at the start of a training course with a simulator. The answer tends to be the same and that answer is Stress Management. For whatever reason Stress Management seems to be the NTS which people do not think they need to know anything about.

Oil and Gas is definitely changing, historically the industry has been regarded as a macho industry "Me man, me tough, Me no want to talk about stress". Thankfully these days/perceptions are changing. If you take nothing else from this book, then please take this: NEVER EVER BOTTLE STRESS UP. It is one of the worst things you can do. The one good thing, in my opinion, to come as a result of the pandemic is the fact that stress and mental health are right up there on people's radars and that is exactly where they should be. I am not sure if there is a direct correlation to the pandemic and stress/mental health awareness on the increase, but all I know is that since the pandemic (COVID) I am witnessing more and more open discussions around these topics. Once upon a time stress might have been seen as weakness and trying to get people, mainly men, in the Oil and Gas Industry to have a chat about stress used to be a non-starter. Now there are open discussions in the classroom, and a lot of the one-to-one feedback is centred around stress and what people can do to try and reduce it. Stress can affect everyone, never bottle it up. I understand the basics of stress and the effect it can have on our cognitive abilities. I am not a trained councillor and do not pretend to be. I can help people to try and process information better to understand what is going on around them, but I would certainly never go further than that during any discussions in the classroom or the one-to-one sessions.

"WHY ARE YOU ONLY EXAMINING THREE NTS?"

This is a question I get asked all the time and it is quite a simple answer and to be honest the most exciting. As I have identified previously, there are six NTS: Stress Management, Communication, Situation Awareness, Decision Making, Teamwork and Leadership. For the purpose of this book and for the TBT training sessions, I focus on the following three: Communication, Situation Awareness and Stress Management. The reason for Communication and Stress Management might already be clear. Communication is the one NTS that people think they know the most about; it is the one skill where we provide most amount of theory and feedback. Stress Management is the one people feel they do not need to know anything about and the last is Situation Awareness. A TBT is all about checking and gauging understanding, and Situation Awareness is about how we take in and process information (Flin et al, 2008). Effective Communication should allow us to check our understanding of the task at hand and also that we understand the environment we work in. Stress can have a negative impact on how we take in and process information; so by covering these three we should be able to increase the effectiveness of the TBT.

Another reason for putting my focus on these three NTS is that the basic concepts for these are something we can discuss at a TBT, how we ask certain questions, how we know the brain is processing information and how we know when someone is getting stressed. However, the others are less likely to be used in a TBT situation. For example, no one is ever going to say what Leadership style are we going to use today? Or which Decision-Making strategy will we be focusing on? All NTS are extremely important when trying to increase safe and effective performance; however, for the TBT I think these three are the most beneficial to learn and apply.

"WHAT DO YOU KNOW ABOUT TBTs?"

Another frequently asked question (FAQ) I get asked, along with "What do you know about offshore, I am assuming you have never worked offshore?" People soon tend to change their tune when I say I worked offshore for 5 years, all around the world, and plenty of that time was spent on the drill floor. Drilling is a completely different beast to anything I have ever seen before (especially in places like Kuwait and Africa); so when you say you have been on the drill floor people tend to take that as Okay and give me a level of acceptance.

Back to the original question of what do I know about TBTs? I like to think I can handle myself with regard to the six NTS. I have created and facilitated numerous NTS training courses across a range of disciplines. I have observed over 100 groups in an electrical simulator; I have observed over 80 teams in a drilling simulator; and I have observed over 50 REPs offshore. I am not entirely sure how many TBT observations I have completed in my life so far, but I would say it could be close to 1000. This number does not include all the TBTs I was part of whilst I worked off-shore. Over the past year, I have presented a series of sessions on improving TBT to well over 750 people, and on the back of that I have observed a further 20 TBTs. Certainly not saying I am an expert, but I like to think it is one area I can really try and help people.

REFERENCES AND FURTHER READING

Cullen, D. 1990. *The Public Inquiry into the Piper Alpha Disaster* (Cm 1310). London: HMSO.
Flin, R.H., O'Connor, P. and Crichton, M., 2008. *Safety at the Sharp End: A Guide to Non-technical Skills*. Aldershot: Ashgate Publishing, Ltd..

SOCIO-TECHNICAL SYSTEMS

Anything by Professor Neville Stanton.
Further reading on Kitty Genovese any Introduction to Psychology. I first read : Gross, R. 2005. *Psychology: The Science of Mind and Behaviour*, 5th edition. London: Hodder Education.

3 Communication

As mentioned in the Introduction, Communication is the one NTS which people think they will excel, both in stressful and normal conditions. However, it is the one we spend most amount of time on for classroom theory and feedback both after simulation exercises and offshore. Get Communication wrong and it could have a huge impact on safe and effective performance (Nieva et al., 1978 as cited in Flin et al., 2008). A frequently used definition of Communication is: "… the exchange of information, feedback or response, ideas and feelings. It provides knowledge, institutes relationships, establishes predictable behaviour patterns maintains attention to the task, and is a management tool" (Kanki and Palmer, 1993). Communication also refers to abilities or styles of interaction while communicating messages. In a small group or team environment, Communication is a means by which tasks can be achieved. Actions are co-ordinated by using instructions, stating intentions, and sending and receiving set information (Wiener et al., 1993).

With regards to the TBT, we use the latter and simplify it down even further and simply say we communicate during the TBT to create a shared understanding between everyone in the group to ensure the group can move forward as one to complete the task.

I have observed numerous TBTs in our electrical simulator, on the odd occasion they decide to do one. I find this interesting, in that they will enter the simulator to start a task yet not conduct a TBT. "Why did you not do a TBT before you started the task?" the most common given answer is that "We didn't think we needed to do one". Would you ever walk into a switch room, start a task and not do a TBT to which they answer "obviously not". So when then did you do it here.

This just emphasises the point that so many people see the TBT as a tick-the-box exercise/something they think they have to do, and it is not ingrained into them. I would like to think if trained pilots were to enter a flight simulator then the first thing they would do is carry out a TBT or equivalent. When we feel we have to do things, the chances of us doing them when people are not around could be reduced. If it were ingrained into people offshore, then the moment they were to walk into the simulator they would do a TBT to ensure they all have an understanding of the task and what it is they are trying to achieve and also give their brain something to focus on.

DOI: 10.1201/9781003493105-3

In total, I think I have observed over 1,000 TBTs. This section is about the main errors or issues I have observed from all my experiences which contribute to ensuring a group does not have a full understanding prior to the job beginning. The main errors are: inappropriate questions, knowing the answer, allow time for questions and feedback, define the term Communication, use of familiarities, get everyone involved and mind reading.

INAPPROPRIATE QUESTIONS

I have seen hundreds of REPs in our training facility so when I go offshore to do TBT sessions and observations there is a very high chance that I will bump into a familiar face. I recently went to a rig where the REP recognised me and said "bloody hell Moffat you have put on some weight, what have you done to yourself?" Thanks for that, he did have a point my survival suit was a little snug; however, it was the 4th January so I was still carrying a little Christmas timber. Now the above question is without a doubt an inappropriate question, however, this is not what I mean by an inappropriate question during a TBT. The way in which you ask certain questions can have huge effect on how you check understanding one of the main reasons we do a TBT.

As an example, if I were to say to someone during the TBT training session, did you understand the definition of the cognitive skills? they tend to answer with yes. However, if I was to ask another person what the exact definition of the cognitive skills? I normally follow up very quickly with it is ok you do not need to answer that. The difference between these questions is one is a closed question, and the other is an open question.

The first one is the closed question, where the person will tend to only answer yes or no. When you ask these questions most of the time you will get the answer yes. People are more likely to answer yes as it is a far more socially acceptable answer, if you answer no then you are opening yourself up to one of two things:

1. Another question being asked: why did you not understand the definition? I gave you a very good explanation. When you answer yes it reduces the chances of another question coming your way.
2. The other reason people tend to answer yes to a closed question is that there is a high chance you could open yourself up to being ridiculed. This is certainly the case for the oil and gas industry.

I remember when I was on a job cleaning tanks and the gaffer said, "Do you all understand?" I said "Erm no I do not understand". How he replied took me a little by surprise, think of the most patronising and belittling voice you can imagine, and he went on to say: "My name is Scott and I'm pretty stupid". He quickly followed this patronising question with: "I'll ask you again Moffat ya big jobbie … Do you understand?" This time I was like "yes, yes I do understand". "Excellent well done that man that was the answer I was looking for". By this stage I was slightly taken aback by (a) the way he spoke to me and (b) the name-calling. I am not sure I had heard that term since primary school.

We answer yes to closed questions, so we do not get another question quickly fired at us and to reduce the chances of us being ridiculed.

Do closed questions check listening? They do to a certain extent; however, for TBTs you are trying to check more than just listening, you are trying to check understanding. Do closed questions check understanding? No, they absolutely do not. Closed questions are no good for TBTs.

What tends to be the questions you ask/hear at the end of or during the TBT? The answers tend to be:

- "Does everyone understand?"
- "Are we all comfortable?"
- "Do we all know what we are doing?"
- "Are we all happy to crack on?"

Do these questions check understanding? No, they do not. When you ask these questions you will probably get the harmonious nodding dog syndrome, a bit like the dog in the old Churchill adverts. These questions are basically a waste of time, as one of the main reasons you carry out a TBT is to check everyone has an understanding of the task they are away to carry out. Getting everyone to simply say yes will not achieve this.

It used to happen all the time when I was on the drill floor. The driller would say "Do you pair of clowns know what you are doing?" "Aye of course we do, you can count on us boss we are going to smash this out of the park". We would leave the dog house and big Kev would say to me "Do you know what you are doing?" and I would say something along the lines of "nah no idea, you?" "Nah me neither, should we go back in and say". We would both look around and the driller would be going absolutely mental at someone else, look at each other and think better of it. Half an hour later, we would have to go in and admit we had no idea what we were doing and face the wrath that we would have gotten before anyway.

There are a lot of people in supervisory positions who have never been trained in the basics of Communication, so when they ask a closed question and do not get the answer they were expecting they can react negatively and throw the toys out of the pram.

I am a shinty player, well player might be a slight/massive exaggeration I am in the park whether I actually play or not is debatable, I have recently been called fat and useless (by a member of my own team I might add), I do have feelings folks.

Shinty is basically like ice hockey on grass: very physical and full-body contact. I was the manager for a number of seasons, I told the girlfriend (now wife) that I had had enough of playing and I was sick of getting battered with balls whilst playing in goals and I also seemed to have developed chocolate hamstrings. I was going to go to the Annual General Meeting (AGM) and retire as I wanted to spend more time with her. She didn't believe that for 1 minute. I attended the AGM with full intention of retiring. After the meeting I got back to the car and called her and asked, "Do you want a cheeky wee McDonalds on the way home?" "What on earth have you done now?" "You are not going to believe this they have only gone and voted me in as manager, I am absolutely delighted and not sure I have ever been so proud in my life as I have never been voted for anything in my life". Her laughter surprised me a little I will not lie, and it was quickly followed by "are you really that gullible?" "No" was my sheepish reply. "Do they have another goalkeeper?" "No". "Do you see what I am getting at here?" another no from myself. "If you retire they do not have another goal-keeper am I right?" Well, I guess it is not exactly the most favourable of positions. "If they do not have another goalkeeper then they will have to find one, however if they keep you on as manager, well I am guessing you would pick yourself, therefore the team still has a goalkeeper". "You see what I am saying now?" "Unfortunately, I do". Needless to say, I did not get her a Mcflurry on the way home.

Coming back to play shinty after the pandemic, like any contact sport whether it be football, rugby, boxing or whatever, was the best feeling in the world. We had basic-ally not been allowed out of the house, well other than a wee walk along the road, and let's be honest that was not exactly the most cathartic thing ever. Going from sitting in the house to full-blown training/playing contact sports was a great feeling. There was a lot of pent-up aggression sitting around in the house and now the government have said we are good to go on the full contact sport front. A friend of mine said when you go back to shinty, or any other sport, after the pandemic I guarantee there will be more sending-offs and injuries than you have seen before. "Why"? Well, no training, no games (just a gentle walk around the block) once you get people, especially your age (thanks for that), going back to play the sport they will be full of enthusiasm. However, as they are not match fit, they will fall like flies: hammies, backs you name it will literally go in the first few seconds. Couple that with grown adults, basically with a weapon (I assume she meant the shinty stick), ready to swing at people, again I assume you mean the ball. However, I get what you are saying. This could be a recipe for disaster.

After lockdown, we were playing a team, they do not exactly like us and they are not on the top of our Christmas card list either. I am in goals as I am the best option, or as the wife would say the only option.

Our full back is a boy called John Moffat, who is useless at shinty, in that he cannot hit a cow's arse with a paddle, but he loves a bit of violence so we take him along on a Saturday afternoon so knock yourself out, a figure of speech: go do what you are good at – hitting people with the stick. In the first couple of minutes of the game, a high looping ball comes over the top and Johns man goes to try and control it in the air. To do this, he lifts the stick above his head, now leaving his ribs completely open. John now sees this as a massive target. He takes the butt of the club and sticks as hard as he can right into the ribs of his opponent. Their player, as good as he is, cannot hit the ball as he is in agony and John canna hit the ball as he is completely useless. You now have two fully grown men belting each other with a stick. That is until John decides to turn around 360 degrees with a huge full swing and hits the player across both knees and he goes down like he has been shot. The referee came running up to John and said: "Do you want to stop doing that?" "No" John replied and walked off. I burst out laughing and the referee said: "Do you think that is funny?" "Yes". The referee is now raging, "You know that is not what I meant you pair of smart arses". Maybe we did but that is the question you asked and therefore the answers you got. It was not John's or my fault for answering with a yes or a no, it was your fault for asking that question; do not get annoyed at us for you asking the question in that way. If you are asking a closed question and do not like the answer you are getting, then do not have a go at the person who answered you. Consciously think about the question you are asking to try and engage other people. Closed questions do not check understanding and are therefore a waste of time for a TBT.

Closed questions do have their uses, for instance when you are looking for short, factual information, "Did you close valve A?" This only requires a yes or no answer you are not interested in where, how or why you closed the valve just purely if it is closed.

Open questions are great for TBTs, as they are a fantastic way of checking understanding, "What was the exact definition of the cognitive skills?" "Yes". What do you mean yes? That does not make any sense. When you ask an open question: what, who, how, why, where etc., then the brain must think and process information which then allows the person asking the question to gauge understanding.

The following are examples of effective open questions which should be asked during/the end of a TBT:

- **What** are you doing?
- **How** are you going to do…?
- **Why** are you doing…?
- **Who** are you away to work with?

Watch yourself with the why question; if you get the tone of it wrong it can come across all wrong and you could really annoy someone. If you annoy someone during the TBT, then the chances of them engaging and participating in the TBT are pretty slim, so trying to check their understanding now is not going to be easy.

However, you get the tone of the why question right and it is a great skill to have in your armoury. You can walk around the park asking people, especially those with less experience, "Why are you doing this?", "Why are you doing that?" and you will soon see how much they will know and, more importantly, how much they do not know.

Get the tone of the question right and you will create two-way conversations; get it wrong and you could find yourself in a bit of a predicament, for instance, "Why are you doing that?" could come across as patronising/aggressive.

We have worked with some drillers from all different backgrounds from around the world, some have been technically brilliant; however, we have come across some who were not the best educated (through absolutely no fault of their own). Some people think that drillers, for want of a better word, are stupid. Trust me, drillers are anything but stupid, they are incredibly switched on. The amount of information they have to take on board, process and then have an understanding of what is going on beneath their feet is actually quite frightening. They are anything but stupid. I do not think there is such a thing as intelligence; I think I might be intelligent at what I do and a tank cleaner is intelligent at what they do. Granted some people, not me, might be great at general knowledge/quizzes etc., but this, in my opinion, does not necessarily make them more intelligent. I have seen some snobbery in my working life, whereby people have spoken down to others just because they deem themselves more intelligent. I am not a fan of this, it takes all sorts, and everyone can contribute in different ways, rant over.

When we teach NTS to some drilling teams we would do a task, whereby we would hand out a piece of paper which had four or five paragraphs. Everyone would read this, then we would ask them questions and we soon find people interpret it differently and give us different answers. That is exactly what is meant to happen, we want to show that even when people are presented with the same information, they can interpret it completely differently. This is all fair and well in a classroom setting; however, this can also happen in the real big bad world. If you have a team who are gathering the same information however are interpreting it differently, you could now have a team who have different pictures in their head, as a result you now have individuals and not a fully functioning team.

I remember once carrying out the above task and thinking that one of the people doing the task might have issues reading. The reason I thought this was this person was what looked like scanning over the page and going at a pace that looked way too quick for normal reading; now they could have been a speed reader, but I was going with that this was not the case.

Once everyone finished reading the facilitator of the course said, "Tell me what you just read" and they replied, rather angrily I might add: "No I don't have to". "I know you do not have to but tell me the main theme you took from this piece". "I have had about as much of this rubbish as I can take". They then stood up kicked back their chair and stormed out of the classroom.

Later, when I was finishing up with my lunch, this person who stormed out of the class came over to me and asked if he could have a quick word for 5 minutes. "No problem at all what's up?" "Well, I am not sure if you noticed but I am not really good at reading?" Notice it was as clear as night and day. "How do you cope offshore?" I know for a fact that at present you are getting at least one procedure, per trip, sent out to you. You have to read this procedure, sign and date it. That document is now a legal document. So my first question is: "Are you reading that document?" "No." "Are you signing it?" "Yes," You can see why I might have a problem with this. "Yes, I can". "Why are you signing it?" "Well, my assistant driller (AD) is really switched on so what he does is he will read the procedure and then tells me what it means and I will try and apply it to my work". "Again, you can see why I have an issue with this, he probably does not have the technical knowledge or experience to have a full understanding of this procedure". So when he is describing to you what it means, it might be somewhat watered down. Not only that, what happens if this person decided I might as well go off and try to be a driller due to the amount of reading/work I am having to do for the driller. "Nah he wouldn't do that as he is pretty loyal". "OK, what happens if they have health issues and have to be removed from the rig?" "Nah he has a good immune system." Ok I can see what is happening here. Right ok let's move on from the procedure, there are numerous other situations offshore where you have to read and process information; for instance, if I were to go into the dog house now and write something on the white board, you would have to read and process that information then make a decision about what is going on beneath your feet. "How do you cope in this situation?" "Well, that is where I might struggle". He then looked at me, winked and said, "It is not as if I am flying a plane or anything dangerous like that is it?". I had no words.

To help some people, like the person before, get a better understanding of the use of open questions, I started using the mnemonic TEDS:

- Tell me
- Explain to me
- Describe to me
- Show me

These are all very useful when trying to get people using open questions, tell me what you think, explain that to the rest of the team, and describe that to me one more time. If you are having a discussion using the first three above and it is clear someone, maybe someone with less experience, does not understand what is being asked of them, then show me is an excellent way of checking understanding. Right ok, you are really not getting this come and show me what it is you are away to do. Don't get me wrong, it does take that bit longer, but you know exactly how much they know and more importantly how much they don't know.

I remember when I was cleaning tanks and we went to a new rig to do some high-pressure jet cleaning. When tank cleaning you could be cleaning anything from huge oil/ballast tasks which basically hold huge volumes of oil and can be the size of a football pitch and some. Normally to get to the bottom of the tank you have to go down about four sets of 20-foot vertical ladders and are in the depths of the tank, trust me when I say it is not the nicest place in the world. However, you can also be cleaning what is known as confined space tanks which are basically storage tanks. They can vary in size but are normally about 20 feet long and 10 feet wide. There is a small entrance hole through which you enter and use a high-pressure gun to try and clean it.

On this particular job, there were two tanks, A and B, about 20 feet apart. It was tank B we were scheduled to clean. Every night after the TBT our gaffer would say to us: "Show me which tank you are working on". We would say this one gaffer, go under our barriers and crack on. He did this without fail every night. After about seven or eight nights when the gaffer said show me which tank are you working on. One of the lads said: "Look I know you are the gaffer and I know you are in charge, but I just want to get this off my chest. Why are you speaking to us like we are kids or we are stupid. I think it is really patronising to ask us the same thing every night, show me what tank you are working on. Sorry but it's doing my head in". To which the gaffer replied: "I am genuinely sorry I have come across as patronising, I certainly do not mean to treat you like kids and I certainly do think you are stupid. We have never worked together or worked on this rig and we are on nights. That could be a recipe for disaster. Please accept my apology, I am just making sure we all understand which tank we are going in before we start". We all think that is a pretty good explanation and then we all enter the tank to start cleaning.

I am not sure if it was true, but apparently when we crew changed there were reports that one team had unintentionally gone in the wrong tank, so they must have moved the "Do not enter" signs/tape and headed in. True or not, but it goes to show when we work on the same thing every night/day our brain can become lazy and see what we are wanting to see (confirmation bias). In the most simple form, confirmation bias is seeing what you expect to see, you have done the task a 100 times, and when you come to do the task for the 101st time your brain has a preconceived picture of what you want to see and not necessarily what you should be seeing, and not what is actually in front of us.

We do TBTs to check understanding across the team, regardless of experience or how many times you have completed a task. TBTs are not questioning technical ability, they are ensuring everyone on that team is familiar with the task and has a similar picture in the brain before they start. Never underestimate the importance of the use of open questions.

Watch yourself, TEDS are very good open questions; however, just by putting the words "can you" in front of them you can take a good open question and turn it into a closed question "Tell me what you think" would become "Can you tell me?", now a yes or no answer. Consciously think about the question you are asking as you do not want to take a good open question and turn it into a closed question just by adding "can you."

LEADING QUESTIONS

We have discussed closed and open questions; another type of question which is hugely important with regards to TBTs are leading questions: leading questions are when you put the answer you were wanting to hear in the question: "It is generator two isn't it"? When you hear the words isn't it (or some variation of this), there is a high chance they have just asked a leading question. It is incredibly difficult for our brains to challenge a leading question as once we hear something, generator 2, it is difficult for you brain to process any further information and, therefore, regurgitate the answer.

You are on day 18, burst/knackered and your brain is like aye it is generator 2. You are away at generator 2, and the supervisor is like wait a minute it's not generator 2. Too late, they already have the cover off and clambering inside. It is incredibly easy to change or manipulate the brain or the mental model by how and when we ask certain questions. TBTs are all about gauging understanding, therefore leading questions are not effective, as the only understanding you are checking is your own.

When I was a lecturer in Psychology I used to say to my students that I could easily change or manipulate your memory/the picture in your brain and they were like: "Shut up no you can't we are like young and cool and that". Aye alright ok.

Memory, in my eyes, is one of the most interesting topics in Psychology and one of the most important. We need memory for everyday functions: if you were to get up now and make a cup of tea or coffee it is memory you are using, if you walk across the street again it is memory we are using. Likewise, when we are working out in the real big bad world we need memory for everything. The brain is incredibly powerful, but my word it can be incredibly lazy and get you into trouble pretty quick.

When teaching memory at the college or high school, memory was seen as one of the interesting subjects you would normally get a good attendance. If the lecture hall could hold 120 people for memory you could have 70–80 students, no problem at all. Not all subjects were like this, I remember turning up to take a lecture on research methods and I think there were three people in the entire lecture hall. About 10 minutes in, two students got up to leave and I was like "Excuse me where are you going?" To which one of them replied "Mate I am proper hanging here and I really don't need this in my life right now" I was a little taken aback; however, I was impressed with their honesty. "This is part of the curriculum and therefore there is a fair chance this will be in your exam, I get this might not be the most riveting of subjects but we will get through it together. Also, you cannot leave this poor person sitting for the next 1 hour and 50 minutes one-to-one with me".

At the end of the 2 hours, I said "So how was that?" Again, their honesty impressed me, one of them said "Mate that was the worst two hours of my life". Good job I have got thick skin.

Back onto memory, you have 70 students in the lecture hall and I am saying I can easily change or manipulate your memory. As it came to the last couple of weeks of teaching memory, I would ask my friend to come into the lecture hall and start having a proper go at me. To which they would say Scott coming in and shouting at you will not be a problem. They would come in, wearing green top and black trousers (or whatever), and start having a proper go at me, "Scott what are you still doing in here it is 12:10 and you were supposed to be out of here by 12". I would jokingly tell her to calm down; now this person was not the sort of person you tell to calm down joking or otherwise. They are now mid-flow: "I do not know who you think you are still being in here this is my time in the lecture hall you arrogant this that and the other, I can see why no one in the staff room likes you" (bit harsh), they would then have a proper go at me. They are now getting the attention of the students and they are probably thinking Scott is a proper tube so this to them is the best day ever. "Wait a minute here, who do you think you are walking in here and speaking to me like that in front of my students? Do you think you are little miss popular in the staff room and if you don't mind leaving and I'll catch up with you later?" We now probably have the attention of the majority of the students, they get a couple more jibes in, and as she leaves she says "that's it I'm going to head of department about this". The moment she leaves, by slamming the door I might add, I turn to one of the students and ask, "What colour of scarf was she wearing?" "Was it blue? Was it black?" Come on, she was just standing right in front of you. I turn to the next person and said, "What about the colour of her jacket, come on you must know the colour of her jacket".

This was not a stressful situation, awkward yes, but certainly not stressful. No one got the answer to the question right unless I asked "What colour of top was she wearing?" or "Was she wearing a green top?" That is how easy it is to change or manipulate the picture we have in our brain. The main reason we are doing TBTs is to create a shared picture/mental model in the brain; this will then allow the team to have a similar understanding of what it is they are trying to achieve. Groups who have a similar picture in the brain can move forward as one and have a better understanding of what is going on around them. Groups who do not have a similar picture in the brain, in real life or in a simulator environment, can separate quite quickly, so you now have x amount of individuals as opposed to one fully functioning team.

Leading questions are not good at checking understanding.

I first read about leading questions in the Loftus and Palmer (1974) (as cited in Gross, 2005) eyewitness testimony study (Baddeley, 1990).

They wanted to show how easy it can be to change or manipulate a mental model based on the questions you ask at a certain time. What they did was they got say five groups and each group had five people in it. Each group would watch a video, the same video, of two cars being involved in a small crash and after they watched they were asked the same question but with a different word in it each time;, for instance, the first group were asked "How fast were the cars going when they bumped into each other", the second group were asked "How fast were the cars going when they contacted each other?", the third group were asked "How fast were the cars going when they collided with each other?", the fourth group were asked "How fast were the cars going when they hit each other?" and finally the last group were asked "How fast were the cars going when they smashed into each other?" (Baddeley, 1990).

The first group those who heard the word bumped gave an average speed of 10 mph, whereas the last group those who heard the word smashed gave an average of 70 mph. Watching the exact same video and asking the exact same question except changing one word in the question you can go from 10 to 70 mph, showing it is incredibly easy to change or manipulate the picture/mental model we have in our brain. Now take this offshore, you are on day 18, physically/mentally tired, and really just want to go home; however, as a supervisor you still need to check understanding. Be careful how you ask questions during a TBT, as you do not want to change or manipulate the pictures others get in their head, especially those with less experience.

KNOWING THE ANSWER

When I worked cleaning tanks, I had a gaffer who would do the same TBT every single day for 21 days, 20 minutes long every day. We knew what he was going to say before he did. We would sit there and listen maybe for the first 2–3 minutes then you would just switch off and start thinking about anything, like I cannot wait to get off of here, maybe treat myself to a wee holiday, Tenerife maybe oh I do like Siam park. I have completely switched off, I am not listening to a single thing this person is saying, I make the assumption the other three or four people are in a similar position (they certainly looked like they had switched off). Towards the end, I might have received a little nudge from the person next to me which meant he was wrapping up so better look slightly interested again. This TBT was a waste of time, no one was listening, which means we have no idea what we are doing, we have different pictures in our brain, and therefore a serious lack of understanding. We are now away to climb inside a tank and not one of us has a clue what the other is doing, not a good start at all.

In short, you have seen something before and therefore assume you know the answer to what is next. Take the Gorillas in the Mist video (Flin et al., 2008; Simons, 1999), it starts and the narrator would say count how many times the team in white throw the basketball. There is also a team dressed in black who are doing the same. The advert video continues, and you begin to count the number of passes the team in white make. At the end of the video, the narrator reveals the answer to be 16 and everyone says, "Yeah i got 16" or whatever number. The narrator will then ask them if anyone saw the gorilla. "What are you talking about? What gorilla?" halfway through the video someone dressed as a gorilla walks onto the screen, does a wee boogey, and then walks off. So many people do not see the gorilla as they are too focused on counting the passes. The video is to show you how much information we miss when we are so focused on something else.

We started using another version of the video (Simons, 2010) where there are two further changes: The curtain in the background changes colour, and one of the team dressed in black walk off the screen. When I show the video, I get the impression like oh here we go we have all seen this video.

But we better humour him, so people "pay attention". I do not think anyone has ever seen either of the two further changes as they think they know the answer. If we think we know what might be coming next, we can switch off very quickly. This can happen very easily when carrying out TBTs; we hear the same person doing the same talk every day, so we simply switch off. It is vitally important we are all switched on before starting any task, so switching off is a no. You need to try and change things up, so people do not anticipate what is coming.

The first way is applying the open questions, go around the group: What are you doing? How are you going to do? When you start asking questions and people are not listening then the next day they probably will be. You turn up the next day and think I better actually listen today as the gaffer was asking me questions and I had no idea what he was saying. Asking questions will focus people's attention, so they can process information and start understanding the task ahead.

Another way to reduce knowing the answer is getting different people to take the TBT; however, make sure they are able to do so. People think taking a TBT is easy (this is certainly not the case); it can be a very daunting task to lead, especially if you have not done one previously. "Right then Moffat, your turn to do the TBT". I remember the first time I heard these words; I was sitting in a room with about 10–20 other teams of people (we would do one big TBT then split into our groups and have a more detailed TBT). At this stage, I hadn't been offshore for too long so I was pretty inexperienced and I had never taken a TBT before, never mind a TBT to this scale. As I sat there thinking there must be another person called Moffat as there was no way he would ever get me to do my first one to all these people, how wrong was I? I stand up, put the piece of paper right in front of my face, and try to start speaking. It was a disaster, I could not get my words out, I was shaking and I was generally all over the place. Some people saw this as an opportunity to have a proper go at me. "This is useless" "Get on with it ya big clown, we do not have all day". The TBT is not a place to ridicule people. People think TBTs are easy, they most certainly are not. They take a lot of practise and can take time to get the confidence to take one, especially to a larger group. Never assume just because people have the technical ability, they can automatically do an effective TBT. I have seen REPs in the smokers who are the life and soul, they are telling jokes and stories, and it appears that nothing seems to phase them. I have then gone out to the park to observe them do a TBT and let's just say it does not go too well; they are sweating and cannot get their words out. I say to the rest of the group "Are their TBTs always like this?" "That was actually a really good one you should see them when you are not observing them". If you are getting different people to read from the same piece of paper it will not be exactly the same; therefore, they will have a better chance of processing the information. Get different people to take the TBT, only if they are competent and confident to do so. Do not throw them under a bus (figuratively speaking).

ALLOWING TIME FOR QUESTIONS AND FEEDBACK

When giving a TBT and asking questions, it is hugely important to allow time for questions and feedback, as it allows to check understanding. What is also important when asking questions is allowing time for the people to process the question being asked. When asking questions allow time for people to process the question being asked; the longer you have been on shift or the more stress you could have been under, then allow more time for the brain to process information (as the amount of information we process week 1, week 2 and week 3 can be completely different).

During simulation exercises and the offshore observations, I have seen two errors people make whilst allowing time for questions and feedback. When someone asks a question, they do not like the moment or two of silence; this tends to be when people are processing the information, so what they do is they either answer their own question or they will ask another question. Both are a waste of time, as for the first one the only person understanding you are checking is your own and for the second you have moved on to another question prior to anyone answering the first. With regards to the first error, people will soon catch onto this; if you continually ask and answer your own question, people will allow you to do so as it makes it easier for them. If we just sit here and wait, he will soon ask a question and answer it himself anyway. Neither check understanding. If you think the question is important enough to ask in the first place, then you must wait for an answer from someone; the silence can sometimes be awkward, but you must see it through and wait. I mean you do not have to stand there in silence for 20 minutes, but certainly enough time for each person to think about the question.

DEFINE WHAT IS MEANT BY THE TERM COMMUNICATION

When observing TBTs I always hear the term Communication being used: "We need to Communicate with each other" and "Make sure we are in constant Communication with each other". Don't just throw the term around willy-nilly. The terms need to be broken down into information that will actually help during the task; for instance, I hear a lot of people use the term two-way verification or simply say we are going to use hand signals. Two-way verification is when one team member relays information to another and receives the exact same information back "1400 psi", which the other person will reply "1400 psi". The reason you get them to say the exact same information back is to ensure their brain has processed the correct information; if they repeat the information yet say "400 psi", then their brain has clearly made a mistake and you can rectify it. I didn't say 400 I said 1400 psi. Information back and forth between parties must be verified. I was taking a TBT session recently and someone said we work in oil and gas so we always verify information; I asked for an example, and this is what they replied: "Valve A is open" and I said how would I reply they said you say "OK". Ok is not a form of two-way verification as by just saying Ok you have no way of knowing what my brain has processed and whether or not it has the correct information. Replying with Ok or anything similar, i.e. aye, roger, copy that etc. are all fine to say after the information has been repeated; however, it must not be used instead. It doesn't always have to be two-way verification, it can increase to three-way (I say it, Anna says it and I say it again) or even four-way verification (I say it, Anna says it, I say it and Anna says it again), depending on the complexity of the task.

If you are ever involved in lone-working, then you no longer have someone to verify information; what I would suggest is what we refer to as self-verification. When working alone say information out loud, i.e. read the procedure or checklist to yourself; if the task is one that does not require the above, then speak aloud. If you are reading out loud or speaking to yourself and you made a mistake, your brain has a higher chance of identifying the mistake. If you are just saying the information in your head, then there is a great chance your brain will miss the information.

I was observing some scaffolders do a TBT. At the end, one walked away, and as he neared the exit he turned as did a hand signal to myself and the other scaffolder. "I have no idea what that means, you are going to have to tell me" I said, "I'm with you I have no idea what that means". After calling him back with the universal hand signal of come here, we had a conversation about if there is a chance you might get separated and you do not have radios or other forms of Communication and you might be in a position where you are using hand signals then you must discuss what each hand signal means. Hand signals are all well and good when everyone knows what they mean and how they are also very easy to misinterpret. "Looks like an international dance off of the YMCA". This was a comment made to our team which consisted of an Italian, an Indian, a Filipino, an Egyptian and myself, when it took us about three times longer to get our equipment together when we were all wearing breathing apparatus due to threat of hydrogen sulphide (H_2S). I am not entirely sure what it does, but I know it is bad.

USE OF FAMILIARITIES

In my opinion, you are more likely to be involved in an incident during routine/ normal work (confirmation bias). When you have completed a job a 100 times, you can become complacent and see what you are expected to see. As a result, be careful of what words you use when giving a TBT. The number of times when doing an off-shore observation I have heard the following phrases:

- "This is a nice easy job".
- "Do not worry this will not take you long, maybe a couple of minutes".
- "You have done this before, in fact not that long ago so you'll be grand".
- "It is just a routine".

The moment you start to use phrases like these is the moment people will switch off and stop processing information. As you have said, it will not take you long at all; they will go out on to the park and think the job will be a doddle. Even though there is information in front of you telling you that is not the case you might not process it as the brain is being lazy as already been told not to bother.

WE CAN READ EACH OTHER'S MINDS

And now for the pièce de resistance, well kinda. "We don't need to communicate with each other, we have been working with each other for years, it's like we know what the other person is thinking, we can read each other's minds". "What colour is Bert thinking about?" "I have no idea" "Exactly you cannot read each other's mind and this whole not communicating with each other as you have worked with each other for years is nonsense." The TBT is to check understanding and also to check that the brain is doing what it is programmed to do. As a result, your TBT is the most important task you will do on a daily basis. I was flying to Houston this year. I checked in and the person at the check-in desk said "Good afternoon Scott, it is good to see you again and thank you for travelling with us again". Yeah, it has been a while since I have travelled anywhere due to COVID but getting back into the swing of things now. "We hope you enjoy your flight, oh just to let you know our pilots have started a new initiative". As they have worked with each other for the last 5 years and done numerous flights with each other we have introduced the new initiative where the pilots do not need to speak to each other for the entire flight. This obviously did not happen; however, if it did, I would say thank you very much, but I have decided not to fly with you guys now or ever. I will take my bags back, please. We would not expect it in any other high-hazard industry, so why would we expect it in the oil and gas industry. We cannot read minds, we have no idea what is going on in the minds of the people we are working with, always communicate, always carry out a TBT, regardless of how often you have worked with each other.

GET EVERYONE INVOLVED

This is one that never ceases to amaze me. "Where is Steve" "Oh he is still in the smokers". "Is there a reason we have started the TBT when all of the work party are not here?"

Make sure everyone who is involved is present for the TBT. Also get everyone involved, including and especially the less experienced. Getting everyone involved can only increase the level of understanding throughout the group.

Be careful of the size of group when doing a TBT. For our simulator training courses we never take more than six people, as once you get over six it can be really difficult to keep the group together. When you get over six the group can start to divide into groups of two, two and three. If you are holding TBT with larger groups, then this can be used as a general overview; once you are back down to your usual groups, do a more detailed TBT. I have seen TBTs with over 15 people; the groups were not working on the same task, and to be honest most people had switched off.

We were not interested in what the scaffolders were doing, and I am pretty sure the feeling was mutual.

COMMUNICATION – MUST DO'S

- Ask open questions.
- Mix it up – get others doing TBTs (only if you know they are competent).
- Allow time for people to process information.
- Define what communication means and what is required.
- Use clear and concise language.
- Get all involved.
- Always do TBT.

REFERENCES AND FURTHER READING

Baddeley, A.D. 1990. *Human Memory: Theory and Practice*. Exeter: Lawrence Erlbaum Associates Ltd.

Flin, R.H., O'Connor, P. and Crichton, M., 2008. *Safety at the Sharp End: A Guide to Non-technical Skills*. Aldershot: Ashgate Publishing Ltd. (Chapter 4)

Kanki, B.G. and Palmer, M.T. 1993. Communication and crew resource management. In E.L. Wiener, B.G. Kanki and R.L. Helmreich (Eds.), *Cockpit Resource Management*. New York: Academic Press.

Simons, D. 1999. *Selective attention test*. www.youtube.com/@DanielSimons, www.youtube.com/watch?v=vJG698U2Mvo. 27 March 2024.

Simons, D. 2010. *The monkey business illusion*. www.youtube.com/@DanielSimons, www.youtube.com/@DanielSimons. 27 March 2024.

Wiener, E.L., Kanki, B.G. and Helmreich, R.L. (Eds.) 1993. *Cockpit Resource Management*. New York: Academic Press.

4   Situation Awareness

"You shower of absolute muppets" or words to that effect is what an HSE lead was shouting at us whilst we were standing in the smokers enjoying a wee inbetweenie. They had every right to be shouting at us, as we had just left the tank we were cleaning to go for this 'well deserved' inbetweenie. We had left the tank and did not tell the Control Room Operator (CRO) that we had in fact left the tank, a big no-no. The picture the CRO has in their head is completely different from the picture we have in our heads. This was exactly why the HSE lead was going ballistic at us.

> I do not mind you guys going for a wee cuppa, you are working hard in horrendous conditions, so have wee break. I am raging at the fact that you did not tell the CRO. They still think you are all still down the tank. Heaven forbid if something went wrong, the CRO has a set of procedures they must work through to ensure you guys get out of the tank safely. Not only that you clowns have turned your radio/s off, so if the CRO is trying to contact you they can't, so their next action would be to maybe send someone along to the tank or down the tank. I am pretty sure it would not have come to that but that is why I am raging, you need to be sharing info to ensure we all know what is happening.

A few sheepish sorrys were then heard throughout the room. One missing or inaccurate piece of information could be enough for two people/groups not to have a similar picture in the brain, which could lead to serious consequences. The TBT will bring people together to ensure they have a shared understanding and a similar picture in the brain.

The more effective the communication, the better overall understanding different people/groups will have. This is why the TBT is so important, especially with routine tasks and/or with groups/individuals who have worked together for a long time.

WHAT IS SITUATION AWARENESS?

In its simplest form situation awareness means "Knowing what is going on around you" (Flin et al, 2008, p. 17). There are more in-depth definitions out there; however, for the purpose of this book this definition is ideal.

    DOI: 10.1201/9781003493105-4

Situation awareness is about how we take in and process information to have a better understanding of our environment. It is to do with perception, attention and memory (Flin et al, 2008). For the purpose of this book, I will focus on memory.

MEMORY

There are two main parts to memory: working memory (WM) and long-term memory (LTM). Basically, when we receive information from our surrounding environment, it comes into the WM and then shoots through to the LTM. I used to ask the question: "What is stored in our LTM?" However due to some of the censored answers I was receiving, I decided to stop asking this question.

Our LTM is basically like a book and on each page of the book a past experience is stored. Information comes in the WM and makes its way through to the LTM. The LTM flicks through the pages of the book until it finds an exact match; it then creates a mental model for that situation. The brain is effectively creating a pattern based on our past experiences. However, if information comes in the WM and through to the LTM and the latter cannot find an exact match, it will most likely start to look for something similar. The picture that has now been created may not be the same as the reality. As our brain controls us it can be incredibly difficult for us to consciously challenge these – mental models, assumptions or pictures we get in our heads (Flin et al, 2008). This is why we encourage effective communication during the TBT and the task itself; the more we communicate and verify, the more likely we are to get a similar picture in our brain and then move forward as one. Effective communication/ TBTs = better situation awareness/knowing what is going on around you.

It is well known that stress can reduce the effectiveness of the WM and therefore decrease situation awareness (Flin et al, 2008). What stress does is puts up a barrier between the WM and LTM, so they no longer speak to each other. If the information we are trying to receive from the WM cannot get past the barrier all the LTM will do is fill in the blanks based on our past experiences without regard for whether they are right or wrong. The more stress we are under, the less information we can process, and therefore we are less likely to have an understanding of what is going on around us; this in turn could mean the more likely we are to be involved in an incident/accident. This also explains why you are more likely to be involved in an incident during routine operations as we have done the task a 100 times when we come to do the task the 101st time the brain cannot process the information as it cannot get past the barrier between the WM and the LTM. This disconnect results in the filling the brain with assumptions which can lead you down the garden path. Our brain is incredibly powerful, but my word it is also incredibly lazy. Regardless of how many times you have done that job or similar, always carry out a TBT to reduce the brain filling in the gaps with assumptions.

How Do We Know the Brain/Working Memory Is Doing What It Is Programmed To Do?

As mentioned in the Communication chapter, open questions are a great way of checking understanding; however, their primary use is to check the WM is doing what it is programmed to do. When the WM is not doing what it is programmed to do then people find it difficult to put specifics in the answer. Say John and myself are away to carry out a high-voltage (HV) isolation, which is not going to end well as I am not technical in the slightest. We head off to start our TBT, if I am asking my open questions: "What are you doing?" "Why are you doing this?" An easy way to tell the WM is no longer doing what it is programmed to do, I might receive an answer like "Just the same as we were doing before lunch".

A friend of a friend of mine has epilepsy and whenever he is going to have a fit they will sit down and look to the right. When they do this, my friends immediately start asking him open questions: "What's my name?" "What's my name?" they tend to get a response like "you know your name". They will then ask another question: "Where are we?" "Where are we?" and might get an answer along the lines of "You know where we are, we have been here loads of times before". They are answering the question; however, they are not able to put any specifics in the answer.

The basics are the same when the WM is no longer doing what it is programmed to do, speak to the LTM (due to that stress barrier). Do not get me wrong if you are carrying out a TBT offshore, they are not going to suddenly sit down and look to the right, that is just what this person did; however, the inability to put specifics in the answer is the same. This is why we ask our open questions to check understanding and also to check whether the WM is doing what it is programmed to do.

The WM can rejig itself very quickly. If I ask Joe an open question and they cannot put a specific in the answer, then I give him a couple of minutes (let him gather his thoughts) and ask someone else a question. This couple of minutes could be enough time for the barrier between the WM and the LTM to start to come down, thus allowing Joe to start to process information and put specifics in his answers. When asking the second or third open question, this should be sufficient time for the WM to rejig itself and start processing information. However, if you are asking 4, 5, or 6 open questions and someone cannot put specifics in the answer, then get them out of that situation.

I recently heard a story about two sparkies. For easy sake, I will refer to them as the older sparky and the younger sparky and they were away to carry out an HV isolation. As they were doing the TBT, the older sparky asked what board they were working on. He got a bit of a fright when the younger sparky said either A or B. Now apparently there were only two boards on that site and they were A or B. The older sparky pushed for answers, but the younger sparky could not specifically say A or B. The older sparky then mentioned that the younger looked a bit agitated. "Right, how about we get out of here and get a wee bit of fresh air". When outside, the older sparky explained to the younger sparky that it was ok he didn't have to say which board it was they were working on, but instead asked what his role throughout the task was. To which the younger sparky said "your role is like safety and things" "Give me an example". However, before the younger sparky had time to think about it, he quickly asked, "What board are we working on?" It became clear that the younger sparky was struggling a little and was not able to answer A or B. The older sparky decided to go for a wee inbetweenie and whilst they were sitting in the smokers the younger sparky said "Can you go and get the medic, I think I need help".

It turned out the younger sparky was struggling a little and ended up getting taken off the rig, by all accounts he is back offshore and doing really well.

"Wouldn't have happened in my day" is what someone said during a training course I was facilitating. "What do you mean?" "No inbetweenies on my rig". I was like what are you talking about this was not an inbetweenie. "Ok, so what would you have done if you went down the stairs when you were offshore and saw these two sparkies sitting in the tea shack". "I would have told them to get their arses back out-side and complete the isolation".

> Well in that case there is a high chance someone could have been injured or worse, the poor lad could barely answer a standard question and you would send him back out to carry on with the job. I am telling you it would not have ended well. What these two did was text book; the older sparky there was some-thing not quite right, he didn't know what it was but got him out of that situ-ation (give him a tin of juice and a sweetie, offshore action for thank you). As for the younger sparky, the honesty and bravery he showed was fantastic, to put his hand up and say I am struggling was a phenomenal effort (give him a tin of juice and a sweetie for the entire next trip).

This is why we ask our open questions to check understanding, and more importantly, to check the WM is doing what it is programmed to do – put specifics in an answer. We should be able to put specifics in an answer after three questions. If you are getting to five or six open questions and someone cannot put specifics in the answer, then get them out of that situation immediately. Never underestimate the importance of the open question.

Most Common Situation Awareness Errors during a TBT: Lack of Detail About What To Do if Something Were To Go Wrong

When you start any task, or indeed after any break (even a short break), it is hugely important to go into detail about what to do if something goes wrong. How often do things go wrong when we are doing our normal daily tasks? Very infrequently; as a result, every time we finish a task our brain is building up a positive picture, positive picture, positive picture. If something were to go wrong, you do not want your brain to automatically go to the positive picture. During the pandemic, I was assessing a couple of REPs and during the HV isolation section, whereby I basically observed the TBT and then the actual task, well the parts I am allowed to observe, and then I gave some feedback. One of the REPs was an older lad who had been there since the begin-ning of time, well except for 2 years he had spent in Singapore and the other lad was about 45 years old; so between them they had a fair few years of experience. When they had finished their task and I was giving them some feedback, I asked "Why did you not go into detail about what to do if something went wrong?" "Well, he has been here his entire career". "Do you think what to do in an emergency is sitting right at the top of his long-term memory?" "Erm no".

The way our body/brain reacts in today's society is the same as when we used to get chased by tigers and bears back in the day. When the brain interprets the situation as stressful, it starts preparing for fight or flight (Flin et al, 2008). Now fight does not mean you are going to kick seven shades out of the supervisor; it means you will tackle the problem head-on, and flight means you will get the hell out of there. There is a third response that could happen, in that you could freeze. Now, you do not want people to freeze when something is going wrong in front of them. The reason people freeze is that the information has gone from the WM to the LTM, if it has got past the stress barrier, and the LTM is flicking through each page of the book/past experience and is like I have no idea what is going on here. I have never seen anything like this so if you hang fire I will get back to you in a minute. We have no idea how long people could freeze, it could be 2 seconds, 2 minutes or 20 minutes (not likely but you get what I mean). We do not want people to freeze, especially when something goes wrong. As a result, even spending a couple of minutes during the initial TBT or after a break going through what to do if something were to go wrong: Where is the isolation switch? Where is the safety hook? etc. This can dramatically reduce the chances of freezing. Also, think about these dynamically, do not just think that is the closest exit so we will go there. You have to discuss your surroundings, especially if something were to change. Make the discussion about what to do if something were to go wrong, i.e. task/environment specific and not generic such as the General Platform Alarm (GPA) goes off then we do this. Process all available information and then make a decision.

A lot of people think working offshore is like going to the fun fair as we get to go on helicopters all the time. As fun as they are for the first trip or maybe two, the novelty soon wears off.

When it comes to a couple of days before home and especially choppers eve (the night before home time), it can feel like you are at a fair ground. However, there are lots of things that can prevent you from going home, i.e. the weather (especially fog). That fun-fair feeling can soon change to that feeling you get when you burn your leg on the helter-skelter or when you are sick on the ghost train. If there is fog around either onshore or offshore, there is a chance the helicopter might not manage to take off or land so you are there until it clears.

Like most people who work offshore, I have been fogged on the platform numerous times. On this one occasion after the first day of getting fogged on, I said to the REP "there is only so much daytime TV I can watch, if I get fogged on again is there anything you want me to do whilst I am here, I could do a procedure or anything just please do not let made me sit through some of those programmes again". "There is nothing really to do, well I might be going out to do a couple of audits if you want to come with me?" "Count me in". I made it clear if I did get fogged on and I did come out to observe a couple of audits I would not be observing him as he had passed his assessment, but I might make a couple of general comments or speak to the elec techs if required.

The next day sure enough I got fogged on and once we were told there was zero chance of the chopper coming in, I asked the REP if I could take him up on his offer. We headed out to the park to a gas module where the techs were working away. The REP asked a couple of questions about the task and then I asked a couple, and my first question was: "why is your gas monitor up here?" my head height (so about 6 foot 1). To which one of them replied, in the most patronising voice you can imagine, "EH in case there is a gas leak". Thank you for that and especially for that tone. "I know the purpose of having the gas monitor but why is it up here?" "Well, there is nowhere else to put it is there? right ok it gets in my way ok so I put it up there". "What gas is it detecting?" "How the hell should I know." "It's H_2S, now the properties of H_2S are that is it heavier than air, and as a result, if you get a gas leak this alarm is never in a million years going to go off if the monitor is up at head height". This is my knowledge of gas well and truly maxed out, I have one piece of gas knowledge and I was able to use it in this situation. "Ok, so a bit of divine intervention and the big man upstairs comes down to help us out and decides to put the gas alarm off". By divine intervention and big man I meant God. To which the other lad replied, "Why would the OIM come down and turn on the gas monitor?" This is going to be a long day. "Right ok, the OIM takes time out of their busy schedule and comes and turns on the gas monitor, where do we go?" The two techs point down the stairs nearest us and said we will head down there. "Why?" "They are the closest stairs to us so we should head down them". "What direction is the wind blowing?" "How the hell should we know". "OK go and check and see if you can tell me which direction the wind is blowing." They both storm off like I have just asked them to do the most unreasonable thing in the world. They come back and say "The wind is generally blowing through this module and towards those stairs because the bright orange flags on the corner of the rig indicate that." "What does that mean?" "Again no idea." It means that if there were a gas leak then the gas would go down the stairs "I hope it held the handrail" (not holding the handrail is a big no–no offshore), now both chuckling away at their joke, which to be fair was pretty funny.

"Ok, the gas has held the handrail and is now making its way down the stairs what direction do we go?" "We go that way" pointing towards the entrance they had just come back from to check the wind direction. "Correct, why?" "Well we have to go upwind, don't we". "That is exactly what I am getting at." One of them turned to me and asked: "How the hell do you know this?" "What do you mean?". "You have only been in here for like 2 minutes, but you knew what to do before we did and we have been in here for ages". "That is my point, the moment I walked through that entrance my first thought was what the hell am I going to do if something were to go wrong."

When we do the same task over and over again, our brain can become lazy and not programme vital information which could result in freezing. Always go into detail in your TBT about what to do if something were to go wrong; it could dramatically reduce the chance of freezing.

This is not some innate information that I am passing on, I still have the words "Moffat, what direction is the wind blowing?" ringing in my ear from when I was working on the drill floor. I was working with a great driller in Kuwait, and he would ask this question every time I would go for a break or come back to the drill floor. This young man might save your life one day; whenever you walk back up here always look at those flags to see what direction the wind is blowing in. That way if something were to go wrong you know you always go up wind, away from the danger. I am still really grateful for that advice; however, I still feel bad for when he asked me how old I thought he was. Now taking into account the heat he had been working in Kuwait and the fact that he was fully exposed to this heat, I thought the wrinkles were age related and not sun related so I went for the late 50s, he was 43.

Now when asked this question I feel it is better to aim low, better to aim low and be inaccurate than to go high and offend.

Once we were allowed back into the simulator following the pandemic, a couple of people were telling me that they think the number of multiple casualties in an incident had increased. I asked why they thought this. Well there were a couple of incidents where someone was injured and instead of thinking about the what-ifs someone else was jumping in to try and help. Now I am no sparky; however, I know that if you jump in to help someone who is already getting a shock you will very quickly become the second casualty. We had not heard of this but can totally see why this might happen. Think about the what-ifs or what to do if something were to go wrong, as it will reduce the chances of you freezing or jumping in two-footed.

Gathering and discussing information before you start from the start, if you take a wee break then have another discussion. You are always trying to build up the most accurate picture of the environment, which can change at any given moment. There is no limit on the number of TBTs or discussions during the task: the more you gather and share information, the more accurate the picture of the environment at that time will be.

I was teaching situation awareness once when this sentence came hurtling at me, "I was on a chopper once that landed on the water and instinctively we did everything we were meant to do. We all got out onto the life raft and waited until we got picked up from a boat. So this thinking about what to do if something goes wrong is nonsense." As this person started going into more detail, it became clear that maybe not everything was done quite as it should have. As he was describing getting into the lifeboat, he mentioned a few people found it harder to get in, as their lifejacket kept getting in the way. Quick as a flash another candidate said "Hang on a minute there were some people who inflated their life jackets in the chopper, well I can see a couple of issues with this. If that thing capsized, I reckon they would have little or no chance of getting out". I am not sure these two people were the best of friends, so now having identified one hole in the story he decided to scratch the surface a little bit more: "How did you get on with the door I heard it is not the easiest to jettison" (in the unlikely event the chopper should land on water you do not open the door in the normal fashion, you should unlock all four corners and push the door out). Looking very proud of himself, he replied "Well I was in charge of that and it was a little hard to slide the door open but once it started moving it was alright". "Mate yer man has a point, if you open the door normally you have now blocked the entire right-hand side of the chopper, i.e. people cannot get out of the windows if it were to capsize. Sounds like you had two things you had to do, jettison the door and inflate the life jacket when on the lifeboat, and by the sounds of it got them both wrong". Chopper man now looking like he was licking pee off a nettle was further annoyed when the person added "I'm calling horse-poop, I am not even sure you were on this chopper I think you just making it up". Making it up or not, I was delighted as it fitted exactly into what we were discussing: always think about what to do if something were to go wrong and gather the information for that current situation. It is not uncommon to get different personalities on a training course and sometimes, like above, they can clash. I always remember a piece of advice my gaffer from cleaning tanks used to say: "If you have two people who do not get on you must address this. They will probably be like rutting stags at some point so you are better to take control of the situation now rather that it comes to a head when you have little control over it". That lunchtime was a fun 30 minutes with the above two, all was sorted though.

All sorted and everyone raring to go, this story got me thinking so I decided to use the basis of the above to try and generate a classroom discussion.

HOW OFTEN DO YOU DO YOUR OFFSHORE SURVIVAL?

The survival Basic Offshore Safety Induction and Emergency Training is a must-do training course for all personnel going offshore. It consists of classroom theory and then application of this theory in certain aspects; i.e. firefighting whereby you enter a metal container and using the techniques provided in the classroom (the hand in front of the face and leg sweep technique) you have to make your way around the container in various scenarios: individual (normal and blindfolded and then as a team). The container only has four corners so if you find yourself walking into a corner with nowhere to go you know you are making a wrong turn. This is exactly what happened the last time I was doing my refresher: the comical sound of someone's hard hat bonking off the side of the metal container did make me chuckle, not so much the instructor. The other exercise is the helicopter escape, which is marmite (in that you will either love it or hate it). I am more on the love side of things, but the three or however many times you do the exercise is about all I can stomach. Basically, you get lowered into the pool, where the water starts to come up. You then have to check the equipment (mouthpiece and direction of window), and once fully submerged you make your way out of the window, once I might add when it is safe to do so, i.e. when it has stopped rotating. No nose clip can prevent water from entering the good old Moffat nose, so there tend to be tears.

Now we do this training every 4 years. So back to the classroom discussion, how many memories has your brain had to process in those 4 years? Thousands. Do you think what to do in an emergency, i.e. the chopper lands on the water and capsizes, is sitting right at the top of your LTM, absolutely not. The only way your brain will be raring to go is if you do your training in the pool and then head straight for the chopper in real life. Obviously, this is not possible; as a result, you will always get shown the video safety brief (prior to mobilisation and demobilisation) to refresh your memory on one of the last things watched/discussed/read prior to the flight. Like all flights, we are advised to re-read the safety card, which has further information on what to do if something were to go wrong, upon settling in on the chopper.

"Going into detail about what to do if something goes wrong is that why pilots tend to give a weather update when landing?" This is a question that was put to me once and, to be honest, I had never really thought about it; however, I liked what they were saying. They went on to say

Well let's think about it, if you are flying anywhere in the world the pilots will come on the tannoy and say something like we are approaching Aberdeen where it is raining, shockingly, and the wind is coming in from the East at whatever speed. They are doing exactly what you are talking about, going into detail about what to do if something goes wrong; however, they are doing it a little more discreet. They can't exactly come over the tannoy and say right if we make an absolute mess of this landing this is what we want you to do. Now everyone on the plane/fixed wing is thinking why on earth would you make a mess of this landing and probably begin to panic. Whereas if they come over the tannoy and say the above, they have told themselves and the stewards the direction of the wind therefore if something were to go wrong during the landing then everyone on the crew knows exactly what side of the plane to take people off and what direction (upwind) to get away from the plane.

I have never thought of this; however, whether it is true or not, I really like this way of thinking. Every time I fly now, I listen out for these instructions and although some are not quite as detailed as what this person said I really do think they have a point.

During your TBT always go into detail about what to do if something were to go wrong; it could reduce the chances of you freezing, and if it is good enough for aviation then it is good enough for everyone else.

Not Discussing Distractions or How To Deal with Distractions

Do you get many distractions at your workplace? I have a funny feeling that the answer is going to be a strong yes. If people are distracted or interrupted while focusing their attention to hold and process information, there is a chance they could forget which steps they have completed (Flin et al, 2008).

I have seen this numerous times in the simulator, the group or individual is focused on one aspect of the task or the simulator and they get distracted. The distraction could be anything between 5 and 30 seconds; some people could take just as long if not longer get back on track where they were before. To be honest, some groups have failed to recover from the distraction. I am making an assumption here; however, if it can take that long in a simulator, where there is little stress present, it could take the same amount of time if not longer in a work environment when there could be more external stressors.

Distractions play havoc with situation awareness, couple that with stress and the environment you are working in then your brain is not processing anywhere the amount of information like it is at this moment, assuming you are reading this in a nicer environment than the workplace. If you even get distracted for let's say 5 seconds, it could take the same amount of time or longer to regroup and start processing information again. As a result, when you get distracted, you must go back to where you were before you were distracted, an absolute must. I know some, if not all, disciplines for certain tasks use either procedures or checklists to aid them, for instance the electrical disciple uses switching programmes when carrying out HV switching and rigging/lifting uses lifting plans when carrying out lifts, obviously. Now when people are using these procedures or checklists, they do not use much WM; as a result, the brain has the ability to follow what is going on and also process further information. These are a great way of slowing the decision-making process down. If you are using these, then the brain can follow exactly where you are and they normally have a column at the side where you can sign your name or tick a box to show completed steps. If you get distracted for even a few minutes, you can look down at your paper and see exactly where you were before the distraction; so nice and easy to retrace your steps. However, not all tasks require a procedure or checklist as some are done from memory or expertise. There are three things I have observed which can happen when you try and go back after a distraction:

- You go back to the exact same place you were before you were distracted (the chances of that happening are pretty slim)
- The second you go back too far, this is ok you can retrace your steps and get an accurate picture in the brain of where you were.
- The third is you do not go back far enough; this is the most dangerous as the brain now fills in the blanks based on what you have seen or done before (confirmation bias – seeing what you expect to see). You could be further ahead from where you thought you were and now have an inaccurate picture in your brain and could go off down a rabbit warren.

When you get distracted, you MUST discuss that you have to go back further than where you thought you were. This is every time you get distracted: if you get distracted at 15:01 then you go back, if you get distracted at 15:31 then you must go back. The information that you were trying to process has not had the time to get to the LTM, and this is why it is hugely important to go back. When discussing distractions during the TBT, it is imperative that you say what you do once you are distracted, i.e. always go back to where you were before you were distracted.

Also, discuss as many (task-relevant) distractions as possible, i.e. the tannoy, other people and what to do if these occur. For example, if the tannoy goes nine times, then leave it if it goes a tenth time and then safely get my attention and we can do something about it. I know distractions offshore occur a lot; as a result, use your TBT to discuss what to do if distracted, go back, and also discuss what distractions could occur during that task, remember they could be different every time you start that task. Do not assume you discussed them earlier, so you are good to go.

To reduce distractions, try and introduce a sterile cockpit; this is an aviation term which prohibits crew members from carrying out non-essential duties or activities while the plane is at a crucial part of the flight, i.e. taxing, take off, initial climb, final approach and landing (Flin et al, 2008). I was carrying out an offshore REP Assessment and during an HV isolation two of the elec techs just started talking about the football the night before: "Did you see that goal last night, offt what a volley, left peg right in the top corner". "Right stop the job, everyone stop and come over here." "Why are you talking about the football?" When you are carrying out a task, any task, you are only discussing the task in hand. You are not discussing anything else. Once the task is complete, then by all means discuss what you wish, but during the TBT or the task itself, you do not discuss anything except the task in hand.

DISTRACTIONS TASK

If you want an easy to highlight the distraction has on our WM, then try this wee memory test. I am sure everyone has done something very similar to this. I first used this when I was a lecturer and I still use it to this day. I use four tasks (can use as many as you feel). Task 1 put individual numbers on the screen (one at a time). Do not use too many; I tend to use 6 for the first task, not too many numbers (just to ease them into it). All those looking at the numbers have to do is to try and remember the numbers on the screen, in the order they appear on the screen not numerical; for instance if they appear 3, 2, 1, then when instructed to do so write 3, 2, 1 not 1, 2, 3. Once all the numbers have appeared on the screen, I will then say start and they have to pick up their pens and write as many of the numbers they can remember in the order they appeared on the screen. It is a memory test not an eye test; as a result, only write the numbers when instructed to do so not when they appear on the screen. Task two is the exact same as task one; however, there are more numbers for them to remember and process, so I tend to use eight. Task three is slightly different, in that I will show them the numbers (same as tasks 1 and 2); however, this time when I say start I do not want them to write anything down. What I want them to do is count backwards, out loud, from 377 in threes, so 377, 374, 371. I will not lie that is probably as far back as I could go. When I say start for the second time, they then pick up their pens and write as many of the numbers (normally eight again) which appeared on the screen. Task four is the same as tasks one and two except this time it is words (eight again) they are trying to remember.

What tends to happen is that for tasks one and two people tend to recall anything from 3 items to 6, 7, 8 (the good old Millers Magic 7 +/− 2 (Baddeley, 1990) (so anywhere between 5 and 9 pieces of information)). However, when we add the distraction in task 3 (normally for around 8–10 seconds), we tend to find that people tend to recall anything from zero to about three, maybe four. We have had one person who remembered seven of the items; however, he did confess that memory was their thing, and they spend their spare time doing things like this, but the norm tends to be relatively low. Once you start the distraction, the counting backwards, then your WM will put its focus on this and not the numbers you were previously trying to remember. As a result, these numbers will not have had the opportunity to go into the LTM; as a result, the chances of them being recalled is slim. This is exactly what happens with distractions offshore/in the workplace: your brain is processing certain information and when the distraction comes along, someone talking to you, the brain will no longer process the old information, i.e. the task and start to process the new information, the conversation, creating an incorrect picture in the brain and this is where your brain can very quickly take you down a garden path. This is why every time you get distracted you must go back to where you were before you were distracted; it will allow your brain to reset and recall where it was, bring the picture back to a more realistic one, and allow you to have a better understanding of the current situation. This 7 +/− 2 tends to be in a nice calm environment not out in the real big bad world. I would be very impressed if your WM could hold more than 3–4 pieces of information when in a stressful environment and also just as impressed if you could hold new information for more than 10 seconds. When we are under stress, the WM does not cope well with processing information, so you must keep communication short, sharp and concise (Flin et al, 2008).

I once watched someone walk into a meeting room turn around and head straight back out the door as they could not remember what it was they went into the room for.

Only to re-enter about 30 seconds later to do the exact same thing, the worst part was they did it for a third time. "For god's sake will someone help that lad out" said a very impatient Offshore Installation Manager (OIM). Once you are getting to the latter stages of the trip, people find it difficult to process information, that is why they open their tally books every minute as they need to write everything down so they do not forget it. Likewise, the amount of yellow sticky notes across people's screens tends to increase significantly towards the latter stages of the trip. You cannot expect people to be processing information all the time; a clear and concise TBT will help people identify and process information which should help them have a better understanding of the task in hand.

IN THEORY, SHOULD YOU REMEMBER/PROCESS MORE NUMBERS OR WORDS?

In theory, it should be easier for you to remember/process the words. The reason is that it is easier to chunk the words together or make an association with them. Chunking is the process whereby you can take several individual pieces of information and combine them to create a singular piece of information. For instance, if I put the words apple, table and car on the screen, those trying to remember could picture an apple on a table in a car. So instead of trying to process one piece of information, each word, they have now chunked them together to create one piece of information. It is so much easier to chunk words than numbers, i.e. 3, 5, 8 mainly as they have less association to each other. Chunking can be a useful skill; however, it can also get humans into a bit of trouble. The example I hear a lot is when driving to the supermarket and once you arrive you have no recollection of how you got there. As you do this route every day, your brain will take in and process individual pieces of information pass them through to the LTM and process them as a whole. However, if you are driving along and farmer Gillies is in the field with his combine harvester and it is out in your periphery, the chances of your brain processing the information is relatively low as it does not support the chuck you are used to processing. As a result, it could dismiss this information and not create an accurate picture in the brain. The same can happen in the workplace if you have someone who has let's say been on the same installation for 10 years, their brain could start to take individual pieces of information and chunk them together, which could result in missing information and the chances of not fully understanding their current situation, thus resulting in an error occurring. Whereas if you take me and put me onto an installation where I have no or little experience or knowledge, my brain will still try and process individual pieces of information and could identify hazards and risks. Regardless of how long you have been on a site, always do a TBT to ensure your brain is processing individual pieces of information and not chucking them together to get the picture you want to see as opposed to what is there in front of you.

Inappropriate 20-Second Scan

Many companies in the oil and gas industry are prompting the 20-second scan. This is where prior to starting work or after a break people are encouraged to search out information which may have changed since they were last there. This has been introduced to try and increase situation awareness; the theory is that if you scan the environment to assess what/if anything has changed, you will process the information, have a better understanding of the environment and thus reduce the chances of something going wrong.

Looking at your watch, and then scanning the environment for that 20 seconds and then saying something along the lines of "Right then folks that is the 20-second scan over, well done us, now let's crack on". This is not a 20-second scan as some people think the important part is the time of 20 seconds; this is not the important part. The reason this 20-second scan is encouraged is to identify any hazards or anomalies not seeing if you can stare at the environment for the said 20 seconds.

The other main issue we see with the 20-second scan is that people just stare aimlessly into the abyss in the hope that a hazard/anomaly will jump out and slap them on the face. If this is what you are doing, you are wasting your time as the brain does not have anything to focus on. When you are doing a 20-second scan give your brain something to focus on, i.e. ask yourself an open question out loud "What is the problem?", "What has changed?", "What anomalies are there in this area?". When you ask these questions to either yourself or the group, you give your brain something to focus on, therefore increasing your understanding of the environment/increasing situation awareness.

We see this in our simulator all the time; on the second exercise there is a piece of equipment that the group has to identify. Without identifying this, they cannot progress with the exercise. When they are looking for this equipment, we hear statements like "There is a problem" and "There must be something we can do" as these are statements harder for the brain to focus, as they do not require an answer. We will then ask a question "What is the problem?" "What is it you are looking for?". As this is a question that will allow their brain to focus, they tend to find the information in about 30 seconds. When doing the 20-second scan ask open questions to allow your brain to focus on the environment and have a higher chance of processing information which could increase situation awareness.

Also during this simulation exercise, we encourage the group to actually go into the switch room and perform 20-second scan. It is very rare that the group will do this; as a result, we go in and change a piece of equipment (I will not go into too much detail in the slim chance that someone attending our training course is reading this book). This is carried out to show that if you do not go and scan your working environment then your brain could be working off a different picture to what is facing you in reality. The more you communicate, scan your environment and process information, the better understanding of your environment you could have.

What I Always Thought Was the Obvious One, but Not Always: Hold a TBT at the Worksite

Another way to increase situation awareness during the TBT, which might seem obvious, is to actually hold the TBT at the worksite. I have observed so many TBTs which are not held at the worksite; they are instead held in the warmth and comfort of the smokers or somewhere else on the asset. Yes, it might be warmer and easier to hold the TBT inside; however, you are reducing the chances of your brain having an accurate picture of the task. Now I get the environment might not allow you to hold the TBT at the work site; however, if this is the case due to it being too noisy etc., then at the very least do a walkthrough to assess if anything has changed prior to the task commencing.

SITUATION AWARENESS – MUST DO'S

- Check the worker is doing what it is programmed to do – ask open questions (can put specifics in their answer).
- Discuss, in detail, what to do if something were to go wrong.
- Discuss distraction and what to do if distracted, i.e. go back to where you were before distracted.
- Discuss 20-second scan (actually ask yourself and the team open questions to engage the brain/identify hazards or anomalies).
- Hold the TBT at the worksite.

REFERENCES AND FURTHER READING

Baddeley, A.D. 1990. *Human Memory: Theory and Practice.* Exeter: Lawrence Erlbaum Associates Ltd.

Endsley, M.R. 2004. *Designing for Situation Awareness: An Approach to User-Centered Design* (2nd ed.). New York: CRC Press.

Flin, R.H., O'Connor, P. and Crichton, M., 2008. *Safety at the Sharp End: A Guide to Non-technical Skills.* Aldershot: Ashgate Publishing, Ltd. (Chapter 2)

Guwande, A. 2010. *The Checklist Manifesto.* New York: Picadur.

Salmon, P.M., Stanton, N.A. and Jenkins, D.P. 2009. *Distributed Situation Awareness: Theory, Measurement and Application to Teamwork* (1st ed.). London: CRC Press.

5 Stress Management

WHAT IS STRESS AND WHY IS IT SO DIFFICULT TO DEFINE?

Stress is so difficult to define as it can affect people completely differently, what has me flapping might not bother anyone else. One thing that we can be sure of is that the oil and gas industry can be extremely stressful, leaving home for up to 3 weeks and working in the middle of the North Sea in a pretty unforgiving environment. The one good thing, if we can call it that, that has come out of the pandemic is the fact that stress and mental health are well and truly on the radars of more people. It no longer has the negative connotations it may have previously had.

I remember having a chat with an older gentleman offshore about stress and anxiety. This person had worked offshore almost his entire working life and he basically turned around and something along the lines of these people, especially young people "just need to pull themselves together, put on their big boy pants and get on with it". Wow really? Thank you for that advice, I will go seek out my big boy pants and I am sure everything will be ok. Unfortunately, this really was the attitude and reality of some people, possibly even the industry, back then, and when I say back then this is not 50–60 years ago this was within the last 10–20 years, at the time of writing (2023). I am not really sure why or how stress suddenly appeared on people's radar. However, during the pandemic from a personal point of view, which might be very similar to others, is that stress was there for everyone to see. I only went offshore a couple of times during the pandemic; however, even during these short trips I could see a huge difference in the amount of stress that people were under. I am not singling the oil and gas industry out here as I am well aware that all industries and individuals are under a huge amount of stress. People do not realise you can get just as stressed doing absolutely nothing as you can when you are going 100 mph all day every day. Now when I say doing absolutely nothing, I do not mean lying on a beach in Tenerife with a pint in your hand. For instance, look at those poor people who were on furlough; through no fault of their own they were sitting at home unable to do anything, after speaking to a few people who were on furlough after the first few weeks/months there was only so much Netflix you could watch. They were not mentally stimulated for 9, 10, 11 months, or however long it was. You cannot then expect these people to walk back into a bank, for instance, and flick a switch on in their brain and not make any errors. Going "back to normal" would have taken a while to get the brain doing

DOI: 10.1201/9781003493105-5

what it is meant to be doing. Likewise, these people were going 110 mph all day every day throughout the pandemic. These amazing people in the National Health Service (NHS) were working tirelessly throughout the pandemic all day every day and not expecting the brain to make errors. I spoke to a couple of friends who work in the NHS and they genuinely could not remember when they had their last full day off. We are not robots we cannot expect to put people under that amount of stress and expect it to not take its toll on the individual, organisation or industry.

I remember working in Mozambique. I was in the middle of the "jungle" not sure if it was an actual jungle but it certainly felt like it. We were about 2 hours away from the nearest main road, right in the thick of it. I was there for 5 weeks and didn't lift a tool, not because I am lazy and bone idol but as a coring engineer if the drilling team do not manage to find the core point (the depth at which the company would like you to start coring) then there is not much we can do. I was checking my equipment every day for the first few days, but even the company person came up to me and said "We are in the middle of nowhere on a secure camp, I am pretty sure your equipment was the same as it was yesterday and the day before that and the day before that". My first thought was Secure camp? There was a village elder, and I mean elder, who would disinterestedly sit near a normal road block barrier, which didn't even cover the entire road, and by that I reckon there was a chance you could have driven a bus through the gap by the end of the barrier and the other side of the road, "oh aye Fort Knox doesn't have a look in here, mate".

Now checking my equipment everyday was off the agenda what was I do to. This was long before Wi-Fi or not sure dial-up was great in the jungle. I still had to get up every day like I was going to work: I would get up at 5 am, get some breakfast then walk from the portacabin to the rig site which crossed about 30 meters of "no man's land", which was between the camp and the rig site. I remember someone saying this was when we were at our most vulnerable of being attacked by the local wildlife. I have no idea if they were joking, but I can safely say I would have given Usain Bolt a run for his money every time I crossed this treacherous path. Once I got to the rig site there was nothing and I mean nothing for me to do. The majority of people there were mainly French speaking, now other than: "*bonjour Je m'appelle Scott, j'ai 16 ans, je vis à Beauly près d'inverness et j'ai une soeur*" I am not exactly sure what I could have said to these Mozambique tigers, replaced the term North Sea Tigers, but I am pretty sure that above would not have helped me in any situation. At least I know my limitations, when we used to go on holiday, anywhere in the world, but especially to Spain, my dad would just put an 'o' at the end of everything he said, and now think he was fluent in that language, i.e. telephoneo, resturanto, water parko. Great effort, dad you smashed it. It is no wonder we, the Brits, are not exactly always welcomed when we go on our travels. When I worked in Kuwait, myself and another lad went to see if we could hire a couple of bikes from the hotel, after what can only be described as the worst game of charades ever, as we left him something along the lines of "Why was I speaking to them like I was basically speaking to someone who worked in local Halfords".

Back to the Mozambique debacle, the first couple of weeks were ok, weeks three and four were really tough as I was not getting much sleep, maybe a couple of hours a night, then still having to leap out of bed like a gazelle and prepare for "work". By the fifth week, I genuinely do not remember being there. I have basically spent a week in a foreign country and I do not remember being there. I have never been so stressed; my brain was like erm nope I absolutely do not want to remember any of that. You can get just as stressed doing nothing as you can when going 100 mph (Flin et al, 2008).

DIFFERENT TYPES OF STRESS

There are two different types of stress: acute and chronic. Acute is when the stress builds up over a short period of time. You see people with this stress all the time off-shore: one minute they are standing there, the next they are throwing their hard hats across the floor. Two minutes later they are totally fine. Chronic is more long-term stress: the budgie died, the dog died and the goldfish isn't looking too good either. This type of stress can take a bit of time to build up and can take some time to reduce; however, with professional help or even speaking to others it can certainly be reduced (Flin et al, 2008).

If you take nothing else from what I say then take this: never ever bottle stress up. It is the worst thing that we can do, apparently, men are especially bad (or good depending on how you look at it) for this. Me man, I don't speak to anyone about my mental health. Men are renowned for bottling stress up; it is the worst thing that you can do even just venting to your friends can reduce stress.

> Mate this pandemic is doing my nut in, I used to go and sit with my pals and we would vent to each other about things like our partners or work etc. and even that felt it was helping. We tried to do it over Zoom during the pandemic, but seeing as it took about 20 minutes to get people signed in and even then all you could see was a close-up of their forehead was not exactly the best way to chat about things, so I felt my stress levels going through the roof.

Never bottle things out and vent to your friends or family, as the saying goes "It's good to talk". I may have taken that out of context, but you know what I mean.

The one we spend the most amount of time on, during any training session, is acute stress, the reason being not many people go offshore and say I am unbelievably stressed, can someone give a cuddle just a wee one nothing major. There are not many, if any, people who work in the oil and gas industry who would do this. Acute stress is a little bit easier to identify and treat with offshore workers. When people know they are getting stressed, they can try and hide it. When people try and hide stress for people who are actually trained it can actually amplify the stress and make it easier for trained people to identify the stress.

Before I went offshore as a Human Factor Director, I heard rumblings that I was being referred to as the "mental doctor" and I was undercover to try and observe as many people as I could then report back to management. I said to the person I heard this from "I take it you corrected them and told them I was there to observe the NTS of their REP". "No way man, this was too good an opportunity to miss. If anything, I probably made it worse, honestly you should see how people are behaving, it is hilarious", Excellent! well I can see where this is going to go.

One of the first nights offshore as a Human Factors Advisor/Director, I was sitting in my cabin and at about 9 pm I hear what can only be described as someone trying to batter the door down. I get up and open the door and a cheery face greets me, "Are you the psychologist?" "Well, no not really, I am here to observe the REP but I am not a psychologist". This did not deter him as the next question near floored me, "Do you mind if I come in and discuss my problems with you?" I look at this guy and I think this must be a wind-up; I look down the corridor to see if I can see anyone sniggering along with the joke. No one there, surely this is not a serious request. I look at him straight in the face and he looks totally serious: "What, do you think I have taken a portable couch out here and you can lie down and we can go all Freudian and talk about all your issues?" "I am not a psychologist, and these discussions are not my areas of expertise. I can have a conversation with you about work etc., but other than that that is about as far as I can go and if this does happen it will be in an office somewhere and certainly not in my cabin".

Believe it or not, this is not the most bizarre question/statement I have ever heard whilst being offshore as the "mental doctor". I was sitting in the smokers once minding my own business, there were about four or five others in the room. One person kept catching my eye, he was looking if I can use the term shifty. He kept looking at me and looking away and was looking rather uncomfortable in his seat. When everyone else got up and left I caught him looking at me again, so I said "Everything alright?" Then without warning he came out with this cracker: "Do you know I have shat myself more times abroad than anyone else I know". "Sorry, did you mean to say that out loud and no funnily enough I did not know that". "Why on earth do you think I needed to hear that?" "Well seeing as it was only you and me left here, I didn't want to sit here in silence, with the mental doctor, as you might have thought I was weird". "So let me get this straight you thought sitting here in silence I might have thought you were weird but you didn't think telling me that you have shat your pants in a foreign country more than anyone else you know would not come across as weird". He looked at me and started laughing, shrugged his shoulders said yeah and left. As the door closed behind him, I could not help but have proper laugh to myself. Being the "mental doctor" did bring some very funny experiences that is for sure.

Another thing this "mental doctor" persona brought was that apparently people would try and change their behaviour around me to try and mask anything they did want me to see. I understand stress a little, again I am certainly no trained counsellor; however, I know enough to identify it in certain situations, especially where stress is present, i.e. offshore. I cannot identify in the classroom or even in the sim (well maybe if people were in there for hours at any one time but not in the 45 minutes we observed them in). The reason is that these are calm and relaxing environments and, as a result, can be easier to hide stress.

When I first went offshore, I used to hide my stress all the time as I thought I was the only one who hated it. I absolutely hated it and I mean hated it; I have to say here the reason I absolutely hated it might have been my own fault. On my first trip offshore, I was wearing my brand-new red boiler suit and was all set to start a new career and earn a bit of money. I was sitting in the smokers offshore, when I used to smoke the one place I never smoked was offshore. The allocated smoking rooms are disgusting; I sat down once to have a fag, looked at the ashtray which was spilling all over the table. I put the fag out and asked anyone if they wanted the 200 I just bought, they went pretty quick. Even though I didn't smoke in the smokers, I always found the smokers to be a better laugh. So, there I was sitting in the smokers listening to everyone taking the mick out of each other and I was chuckling away to myself thinking I quite like this atmosphere I could get used to this. As I turned around, I accidentally caught the eye of one of the scaffolders. I didn't know at the time he was a scaff but I was away to learn this and another couple of things very quickly. As I made eye contact with the scaffolder, he said to me "Do you want my autograph?" As he said this, I remembered my one and only comeback (my friend Steve had told me his brother used it once) as I sat I thought what a perfect time to use it: "I would be surprised if you could write" left my mouth and I remember instantly regretting it. The room made a noise like the air was being sucked out of it and there were a couple of people who burst out laughing and said "ooooooo he has proper done you there mate"; this really wasn't helping my cause in any way. I remember the person next to me saying "What are you doing? If you want my advice keep that big mouth of yours shut" – information I feel I could have done with 2 minutes prior to my outburst. Before I could say anything the scaff was off on a roll, and he absolutely bombarded me with insults, he completely destroyed me and I had nothing. I had already used my one comeback and all I could do was sit there and get verbally destroyed. I would love to say it ended there but sadly it didn't. Every time I saw the scaff he would destroy me (and I mean destroy me), whether it was when I walked past him in the morning I would say alright and he would carry on where he left before or it could have been in the galley in front of x amount of people it didn't matter if he saw an opportunity he would take it. The worst part was I was on the same chopper as him on the way home; therefore when I was to return for my second trip, I was on the same chopper on the way back offshore. When I entered Bristow, one of the helicopter terminals in Aberdeen, I caught sight of him in the distance, well I will be staying away from there. I went for a sausage roll and took a wee sachet of tomato sauce. As I left the café who did I nearly bump into; of course, it had to be him and in front of everyone at Bristow's he destroyed me there and then. As I walked back to my seat, I thought well at least my day could not get any worse, unfortunately as I went to take my first bite of my sausage roll a blob of tomato sauce fell out and dribbled down my jumper. It was not white, thankfully, but it was pretty clear I had dribbled something down my top. The next 3 weeks were very similar to the first three, any opportunity to rip into me the scaff took it. As this insult barrage continued over the 3 weeks, I was really beginning to hate my new offshore experience. I go home for 3 weeks, and then after about 17 days I am now really not looking forward to heading back offshore. I get back on the asset and everything continues as per normal. Halfway through this trip myself and the lad I was working with decided to go for a wee inbetweenie, I'll head in to make the cuppas and meet you in there. As I walk into the smokers, who is sitting there the scaff and one of his pals. He doesn't really take any notice as he is engrossed in bending the ear of his pal. The thing about this scaff is that he had pretty bad teeth, in that you could stick a broom in the gaps of his teeth. As I am standing there pouring the first cup of tea, I hear him say to his pal "Do you know how I finally managed to get the mrs to go out on a date with me?" Now without thinking I perked up with "Well it wasn't your cheeky wee smile was it?", "What the hell did you just say?" Now I am in trouble, "Have you got a death wish?" Without turning I pour my tea down the sink and I do a side-to-side shimmy towards the door then slink out into the corridor. I walk about 10 yards and I hear the door behind me close; I turn round but I already know who it is. As I turn, I am now directly facing him, he said something along the lines of "Don't you ever speak to me like that", and before I know it he barges me with his shoulder and walks on. "What is your problem man?" – yet another incident my mouth acts without the permission from my brain. I then follow up with "You know what I am not interested, whatever." Great argument there Scott. Now slowly but surely the mick-taking began to slow down and eventually it stopped. I am not saying that my talking back to the scaff had anything to do with this; I think he just got bored and moved on, took him long enough. Now the reason I mention this is when I was going through these unpleasant weeks I was sitting in the smokers or my room and thinking I was the only one who hated offshore; now when I go offshore these days and sit in the smokers I can see stress is present amongst a lot of people. All I am aiming to do here is to give the basic tools to try and identify stress in yourself and others. The TBT is the perfect opportunity to do this. I have heard people say what about those people who try and hide it; the more people try and hide the stress from others, especially if people are trained to identify it, the easier it is for them to identify the stress.

THERE ARE FOUR MAIN WAYS TO IDENTIFY ACUTE STRESS

BEST is the four main ways to identify acute stress: Behaviour, Emotion, Somatic and Thinking.

BEHAVIOUR

When I ask the question what typically happens to our behaviour when we get put under acute stress, most people will say we get aggressive/irritable/irate. Now we have probably described about 90% of the people who work offshore. So if I am always shouting, screaming and balling, what typically would happen is that my behaviour would change; it will most likely go the opposite way to how you normally behave (if you are quiet you will become more irate/aggressive, and if you tend to shout scream and ball then you will tend to go quiet). This again is why the use of open questions is so important. If you are doing a TBT and someone's behaviour changes, then you must ask a quick open question: "What are we doing." That way if they can put specific information into the answer, you know the working and LTM are speaking to each other, they can process information and they are good to go. If they are unable to put specific information into the answer, then ask another couple of questions until they can; if they cannot, then remove them from the situation. Another aspect that tends to change when we get put under acute stress is the speed of our speech, in that it tends to go through the roof, as does the tone. We were involved in an incident investigation review and an electrical technician was asked "What happened right before your mate electrocuted himself?" Well he started speaking incredibly fast and well what did you do? "Nothing, I thought it was pretty funny". There will always be a change in behaviour; however, like the example given above, if this change is identified, then something must be done about it, as identifying the change and then not doing anything could really sit with someone for a long time. During the TBT or the part where it says HF do not just tick the box, say if my behaviour changes at all then stop the job and wait until you are, I can answer open questions correctly (i.e. check the brain is processing information). Discussing these things at the TBT is important; however, so is the acting on it – if you see something you must process it and act.

One of the things we also do a lot of work on is incident investigations, with the help of Prof. Neville Stanton; I like to think we are not too bad at these. When doing an investigation we try and be part of every aspect including the interviews with those involved. In the odd chance we get to speak to those who were involved, directly or indirectly, and just having general conversations with people offshore, we tend to hear a similar thing if the incident involved stress and that tends to be the person involved just was not themselves. Again, this highlights that when people are under stress their behaviour will always change.

Of all the examples or stories I have provided so far this is the only one I will share where I am not entirely sure is true or at least 100% true, even so for this section it does give me a great example.

Now either I have developed x-ray vision or that man needs some serious help, said one offshore worker whilst standing in the control room, along with 15–20 others, whilst everyone was going through their tasks for the day. He was referring to the fact that someone had entered the room wearing their trainers, a pair of their finest brown y-fronts and either a t-shirt or Rab C Nesbit vest. It did not stop there; they stood in the middle of the room, proceeded to pull their pants up as high as they could then spun around on one leg. The medic soon realised something was not quite right, so got everyone out to deal with this person. Now what is perceived as 'normal' behaviour to turning up at a morning meeting wearing your undies. There would have been plenty of opportunities to see the change in this person's behaviour. However, due to lack of knowledge/training I fear these behavioural changes were unfortunately missed. This is why we believe that all offshore personnel should be trained in the basics of stress management and stress identification to ensure incidents like these do not get to the stage.

The part I have trouble believing is during the investigation everyone who was in the room at the time were asked: "What did you see that morning?" This question was in regard to the undie-wearing person's behaviour; however, true or not, apparently one person replied with: "I could clearly see a bit of hairy ball". Excellent, I am sure some people, including myself, had a wee chortle at this. Although this is humorous, the underlying issues are stress and stress identification certainly are anything but funny.

"Have either of you boys seen my dog?" This sentence was uttered to a couple of people offshore. It was about the end of January nearing 5 o'clock in the evening; it was already dark as an elderly lad approached two younger lads and said the above sentence. Now instead of thinking we are in the middle of the North Sea, why on earth would you be looking for your dog. The penny should have dropped that this individual might need some help. What made it worse was as they approached the two younger lads one made a comment to the other along the lines of Joe Bloggs does not look too good. So that coupled with him now asking if they have seen his dog alarm bells should have been ringing. However, instead of putting two and two together, they made a joke like aye it is over there near that container.

Everyone goes to bed and no one thinks anything of it. The next morning the REP was having a general chat with the team and said "Where is Joe Bloggs?" "Anyone seen Joe Bloggs?" At this stage, one of the lads said "I have not seen him since before tea last night". What was he doing? At this point, the other lad started laughing and said "Well he was looking for his dog", and the other said "now I think about it he didn't look too good either. He was kinda hunched over and looked a bit peely-wally." Did either of you clowns go to the medic or at least tell anyone? The atmosphere changed a little in the room from a little light-hearted chat to a bit more serious. Right, you pair go to Joe's room, you two to the medic and you with me. After about 10 minutes of frantic searching with no luck. The REP decided to take a shortcut through an HV switch room and got the shock of his life. Here was Joe standing there wearing his trainers, a pair of shorts, and a hoodie. I know I do not go offshore much these days, but I know this is not the correct personal protective equipment. By all accounts, the REP did a fantastic job in getting Joe to the smokers and then getting the medic down to help.

When dealing with people who are under stress, rational thoughts can quickly turn irrational and human beings can turn freakishly strong incredibly quick. When dealing with people who might be suffering from stress the communication is paramount. All you are trying to do is reduce the anxiety levels, so if you are walking about offshore and you come across someone who you suspect is suffering from stress/anxiety, i.e. they are walking about saying things that might be out of character, be careful how you approach and more importantly how you speak to them. If someone is walking about saying "See yer man Scott I absolutely hate him he is an absolute tube", the one thing you need to do is agree with them; do not disagree with them. Do not say "What are you on about Scott is a great guy it is you who has the problem, look at you, you got snot and saliva all over you". If you disagree with them or say something like that, then you might find yourself in a bit of trouble quite quickly. The important part is to remember to agree with everything they are saying so with the example above just say "Yeah you are right, he is a tube". What you will find is they might go on to another topic, but again regardless of this topic change always agree with them. They may change the topic a good few times or even go back over an old one; whatever they say always agree with what they are saying, and all you are trying to do is get their anxiety down. If you are trying to get them to do something, then try and use positive language; if you want them to do something, then do not put the negative in the question or statement, i.e. do not sit down as they will more than likely sit down. The state the brain is in it is not fully processing information and, as a result, could miss information which is most likely going to be the do-not section of the sentence.

EMOTION

With regards to emotions, it does not mean I feel happy I feel sad. I remember one of my friends telling me a story: when they were at a swimming meet and there was a swimmer who was walking about and saying that they could not do a tumble turn, now I can do a tumble turn not very well but I will give it a bash (I somehow end up facing the same way I came). Now here is a swimmer who has been training their entire life, is a well-experienced masters swimmer, and walks about saying they cannot do a tumble turn. The emotions that we experience when under acute stress is that it makes us challenge our own, confidence, competence and ability. Now the swimmer was not walking about shouting at the top of their voice: "Oh my god I canna remember how to do a tumble turn". They were just walking about saying under their breath that they could not do tumble turn. Now there might have been a couple of competitors thinking this is fantastic we got this, she is wondering around unable saying she canna do a tumble turn, and even has her swimsuit on the wrong way we got this in the bag. One of the swimmers left and went and got her friend or coach and basically said I think you need to go and see your friend downstairs, I think she is in a little bit of bother; so they went to help calm her down. This is why you need to listen, especially during a TBT, if you have someone saying look I am not sure I can do this today. Do not react by saying something like what is your problem, you have been doing this for years now get out there and get cracking. What you are trying to do is reduce the acute stress/anxiety. Start by asking what is different from yesterday. Again, once they can put specifics in the answer they are good to go. Like the swimmer they will not be shouting it from the rooftops, so listening to what everyone is saying during the TBT is incredibly important. You do not want someone to be saying this; it is not picked up and you now have someone away out to the park who is genuinely challenging their own abilities, confidence and competence. It is an accident waiting to happen. Always listen to what people are saying, especially if it is something negative or unusual.

SOMATIC (PHYSICAL)

Somatic, basically means physical. I am guessing; whoever, came up with the acronym BEST, they thought it was easier to remember BEST than BEPT.

There is one easy way to spot acute stress in people and that is … sweating. As we have already mentioned, men cannot be great at opening up about this. People, especially men, do not get off the chopper when they arrive back offshore and say I am unbelievably stressed, who wants to give me a cuddle, go on just a wee one nothing serious. People tend to hide it, so what the body does is it is like I am struggling a little here. So instead of asking for help or a cuddle I know what I will do I will do something subtly so others know I am struggling. So what it does is opens up the pores and people sweat. When I say sweat I do not mean a little bead of sweat trickling down the forehead, I mean you look like you just came out of the shower. It has happened to me a lot, especially when I was working in foreign countries by myself and I didn't have a clue what I was doing: Company man "Do you know how much money this is costing us?" "Yes, actually I do but asking me this question every 2 minutes is not really helping me to be honest".

When we assess REP offshore, we prefer for the REPs to do the 5-day training course and then we go offshore to observe them in real life, using the feedback from the simulator as a basis. Do the classroom theory, get observed in the simulator, get feedback from the simulator, and go offshore to observe the theory and simulator in real life. To us that makes sense. However, we have, over the years, had a couple of clients who insisted we go offshore to observe them first then they do the 5-day REP training course. Nothing like putting the cart before the horse; we argued that there was no point in this as we would have nothing to observe as they had not heard any of the theory. We ended up doing it once to show that it really didn't work. Whilst I was offshore, I observed the TBT for an HV isolation and then the task. Once the sparkies had finished they called me back into the room. When doing an HV isolation those involved must, depending on company-specific procedures, wear what is known as an arc flash suit, basically look like the marshmallow man from the original *Ghostbuster* film. In this instance, the HV isolation was complete and one of the sparkies took off the arc flash hood and he was absolutely dripping in sweat. In a flash, no pun intended, the other sparky says "Wow man look at the state of you, you are ringing". As I stand there looking at him, I think to myself I know what is up here. The one thing with sweat is once you start it can be very difficult to stop, especially if people keep highlighting it. When we were sitting in the smokers, the poor lad was just sitting there drenched in sweat and it only got worse as every time someone walked in and saw him the inevitable words of "bloody hell man look at the state of you" could be heard. I was like just take yourself out of this situation and go cool down. When we are offshore, we observe the REP doing everything, so it is very rare we are ever more than 3 feet from them. As we left the smokers I accidentally followed the wrong person, so instead of following the REP, I am now following the sweat lad. As I followed him into a small room, he said "When did you know". Well, I wasn't entirely sure until about 2 seconds ago when you asked, but what drew my attention to you was the sweating when you took off your arc flash hood. We sat down and had a conversation and I gave him a couple of pointers. Like behavioural change, sweating is a great way to identify acute stress.

Like I say it has happened to me on numerous occasions. When I was younger I sometimes thought I was going have a massive panic attack and drown in my own sweat when I was walking through the Eastgate shopping centre in Inverness, which for reference is not exactly big.

"Ah ok like Alan Sugar", if what was said to me during a one-to-one conversation between a senior manager and myself during my teaching days. This came about when I was sitting in their office, she was directly in front of me and between us was a desk. The door was to my right and the outer wall was to my left . Now this person had a reputation with both the students and staff as one not to cross; they came across as a strong independent person and didn't really take crap from anyone. So sitting face to face with them in what I would describe as an already pretty warm room, where I was feeling a little uncomfortable and thought I could break out in sweat at any moment; however, to her credit they were engaging me and asking me easy enough question like "How are you getting on here at the school?", "How are your students?" They went on until she said "Are you here on Friday?" to which I replied I no longer work Fridays, the previous year I would do 2 hours on the Tuesday and 2 hours on the Friday at the school. To which I replied, "I no longer work on Fridays, I now work part-time back in Aberdeen and went on to describe People Factor Consultants to her about how the owners had started up their own company and what they did". As I progressed, she looked pretty interested and it was here I made a massive error: She proceeded to say "Ah like Alan Sugar" I am assuming like business/entrepreneurial people, who have not only created their own business but now successfully running it. Now I am not sure if I was not paying attention or whether sitting in this hot, stuffy room was not beginning to get to me. But I did not hear "Ah like Alan Sugar" in my head, I heard, "Ah ok sugar" and I will not lie for a brief moment I thought did she just call me sugar, that is a bit weird. I never thought anything of it, so I replied: "That's right honey". Silence, she is now just looking at me like a confused dog and said something along the lines. "What? Did you just call me honey?" "No what are you talking about?" was my sheepish answer. "I am pretty sure you just called me honey". At this stage, I got really nervous and let a nervous grin get the better of me, oh no please not now. "Do you think this is funny? Honestly I do not think this is funny". "I am sorry I thought you just made a joke and called me sugar, so I replied honey". Thinking this feeble answer would solve everything and we would just carry on. This could not have been further from the truth. "I am sorry I really do not think why you think you can sit here in my office and call me honey and more importantly why do you think this is funny, you have crossed a professional line Mr Moffat" Oh no it's now Mr Moffat and not Scott. I am now in a world of pain. "Is this some sort of joke either the staff or students are playing because if it is then I really really do not think it is funny, in fact, funny is so far off my radar right now". All I could really muster was a faint "I am really sorry, I think I should go". As I stand to leave, I am now looking right at them, and I can tell she is just as raging as she is confused. Now I know the door is on my right-hand side; however, when I stand to leave, I pick up my bag like a little boy holding it for dear life across my chest and I proceed to turn to my left, where I am greeted with a massive outer stone wall. As I stand, staring at the wall I now realise exactly what I have done and also I can feel my body, especially my face, really heating up and then it started: sweating like an absolute mad man, it was running off my face, it was like I just came out of the shower. As I turned around and sat back down, still clutching my bag like a boy. They asked if I was ok to which I replied I am not good at all. "Do you want to leave?" "Yes please". As I opened the door there were two of my students standing there looking at me and one of them said: "Bloody hell Mr Moffat you look awful, are you ok?" I couldn't even muster a reply and walked off down the corridor. As I got about halfway down the corridor, I could hear the manager say "Mr Moffat it is ok I now realise where the confusion has occurred". I could not even turn around never mind speak to them, so I basically put my hand in the air and mumbled sorry and carried on walking. When I got back to the college, I explained to my friend what had happened and she said, "Even if the manager did call you sugar at what stage, you big idiot, did you think it was acceptable to call her honey?" There was no real communication between the manager and myself over the next couple of weeks, not that she was ignoring me but there was no need for any communication and I certainly was not going to start any. When she finally did communicate, it was after a meeting with other colleagues; as she walked past, she said something along the lines "no words of endearment for me today?" I decided it was best to stay quiet and smile politely.

THINKING

With regards to thinking, acute stress tends to affect WM, decision making and prioritising. People find it very difficult to focus on one task. If you are sharing an office with someone who is sitting there typing away on an email, then they might get up and start making a cup of tea, and then might be over faffing around the printer. Their brain finds it difficult to focus on one task, so it is here there and everywhere. Or they might find it difficult to focus on the task at hand. When working down in the tank, when on the gun, which could be PSI as high as up to 20,000 PSI (or even higher). This will easily hurt you very badly. When down the tank you wear protective trousers which are basically stuffed with padding to prevent from hurting yourself. The other safety precaution is that if you were to slip and take your hands off both triggers, it could take up to 30 seconds to go back down to zero pressure, which could still do some damage to you, someone else or equipment. As a result, you will have someone behind you in an emergency; so if you slip and both hands come off the triggers they immediately hit the kill switch to stop the water coming out of the gun. When you are on the emergency switch, you are there for the entire duration; this is not negotiable.

I remember being down the tank and the tank entry person kept getting me to stop jetting, so I would take my hands off both triggers, wait for the pressure to go to zero then I would turn around to see what the issue was, and lo and behold the lad who was meant to be sitting by the emergency stop button was nowhere to be seen. He was about 30 feet away sweeping the floor. Now I didn't know it at the time, and still not 100% sure to be honest, but looking back it could have easily been stress, in that he could not sit there for 2 hours at a time with just his own thoughts. As a result, he had to get up and do something.

BEST is a great way to identify stress, and it should be incorporated into every TBT. Identify the stress then ask a couple of open questions, if they are able to put specifics in the answer then they are good to go; if not, then ask another question or two, and if they are still unable to put specifics in the answer, then stop the job. Please do not think someone has to be showing all of the above, i.e. if someone's behaviour has changed, however they are not sweating like a mad man, then do not think this is not stress. These are the most prevalent in people; it can be any one of these at any one time and not them all at once. People do not have to be showing all of these. Put these in your LTM and always discuss during your TBT to try and identify stress and ensure the brain is doing what it is programmed to do.

STRESS MANAGEMENT – MUST DO'S

- Use BEST to identify stress in others and yourself
- More for life in general – NEVER EVER BOTTLE STRESS UP!!!!!

REFERENCES AND FURTHER READING

Flin, R.H., O'Connor, P. and Crichton, M., 2008. *Safety at the Sharp End: A Guide to Non-technical Skills*. Aldershot: Ashgate Publishing, Ltd. (Chapter 7)

Palmer, S. and Cooper, C. 2013. *How to Deal with Stress* (Vol. 143). London: Kogan Page Publishers.

Stanton, N.A., Salmon, P.M., Rafferty, L.A., Walker, G.H., Baber, C. and Jenkins, D.P. 2013. *Human Factors Methods: A Practical Guide for Engineering and Design* (2nd ed.). Boca Raton, FL: CRC Press.

6 Summary of All Key Points from All Sections

Using the following key points should allow everyone of all abilities to perform more engaging TBTs which should lead to safe and effective performance at the work site. Communication

- Ask open questions.
- Mix it up – get others doing TBTs (only if you know they are competent).
- Allow time for people to process information.
- Define what communication means and what is required.
- Use clear and concise language.
- Get all involved.
- Always do TBT.

Situation Awareness

- Check the working memory is doing what it is programmed to do – ask open questions (can put specifics in their answer).
- Discuss, in detail, what to do if something were to go wrong.
- Discuss distraction and what to do if distracted, i.e. go back to where you were before distracted.
- Discuss 20-second scan (actually ask yourself and team open questions to engage the brain/identify hazards or anomalies).
- Hold the TBT at the worksite.

Stress Management

- Use BEST to identify stress in others and yourself

DOI: 10.1201/9781003493105-6

TBT Checklist

TBT is at the worksite (or a suitable replacement if environment does not allow at worksite) ☐

All are present and involved ☐

Use open questions (What, Who, Where, Why, Tell me, Explain to me, Describe to me + Show me)☐

All have answered a question/able to put specifics in the answer ☐

Allows time for questions and answers ☐

Define what communication is required and who is communicating with who ☐

Use clear and concise language (not this is easy or this is just a routine) ☐

Discuss, in detail, what to do if something were to go wrong ☐

Discuss distractions and what to do if distracted, i.e. go back to where you were before distracted ☐

Discuss 20-second scan (actually ask yourself and team open questions to identify hazards) ☐

Discuss Behavioural, Emotion, Somatic +Thinking and what to do if these are observed ☐

FIGURE 2   The Toolbox Talk checklist.

Index

A

Acute, 50, 53, 56–57, 59
Allowing time for questions and feedback, 1, 30
Assumptions, 35

B

Behaviour, 51, 53–54, 59

C

Care assistant, 2
Checklist, 31, 43, 47, 62
Chronic, 50
Closed questions, 20–21, 23, 26
Cognitive skills, 13, 20, 23
Communication, 61
Confirmation bias, 26, 31, 43
Coring engineer, 3, 4, 49
Crichton, M., 18, 33, 47, 60

D

Define what is meant by the term communication, 20, 30, 33, 61
Definition, 13, 19, 34, 48
Director, 1, 9, 50
Distractions, 42–43
Distraction task, 43–45
Dog house, 24
Drilling, 5, 17

E

Electrical, 17, 19, 42, 53
Emotion, 53, 55–56
Errors, 37

F

Fight, 37
Figure, 14
Flight, 37–38
Flin, R. 13, 15–19, 28, 33–35, 37, 42–44, 47, 50, 60
Freeze, 38

G

Get everyone involved, 1, 20, 32–33, 61
Gorillas in the mist video, 28

I

Inappropriate questions, 1, 20
Inappropriate 20 second scan, 45

K

Knowing the answer, 1, 10, 20, 28–29

L

Leading questions, 26–28
Lecturer, 5
Listening, 4, 6, 15, 21, 28–29, 52, 56
Long term memory (LTM), 35–36, 38, 41, 43–45, 53, 59

M

Memory, 5, 26–27, 33, 35, 37, 41, 43–44, 47, 61
Mental models, 35
Millers magic number seven, 44
Must do's, 33, 47, 59

N

Non-technical skills, 1

O

O'Connor, P., 18, 33, 47, 60
Offshore, 2
Offshore survival (BOSIET), 41
Open questions, 23, 25–26, 29, 33, 35–37, 46–47, 53–54, 59, 61

P

Pattern, 35
Personal resource management, 13

R

Rotation (offshore), 3

S

Self-reflection, 14
Shinty, 21–22
Simulator, 5, 9, 14–17, 19, 27, 32, 40, 42, 46, 56–57
Situation awareness, 61
Social skills, 13
Somatic, 53, 56
Stress, 35–38, 42, 44
Stress management, 61

Style of writing, 11
Summary, 61–62

T

Tank cleaner, 2–4, 9, 20, 23, 25, 28, 34, 40, 59
TBT at the worksite, 46–47, 61
TEDS, 25–26
Thinking, 53–54, 56, 58–59

U

University, 1–2
Use of familiarities, 1, 20, 31

V

Voluntary firefighter, 1

W

We can read each other's minds, 32
What to do if something went wrong?, 37
Who am I?, 1
Why I wrote the book?, 11
Why only write about three non-technical
    skills, 17
Work experience, 1, 9
Working memory (WM), 35–38, 42–44, 53

Printed in the United States
by Baker & Taylor Publisher Services